Jamal Berakdar, Jürgen Kirschner (Eds.)

Correlation Spectroscopy of Surfaces, Thin Films, and Nanostructures

Edited by
Jamal Berakdar, Jürgen Kirschner

Correlation Spectroscopy of Surfaces, Thin Films, and Nanostructures

WILEY-VCH

WILEY-VCH Verlag GmbH & Co. KGaA

Editors

Jamal Berakdar
Max-Planck-Intitut für Mikrostrukturphysik
Halle, Germany
e-mail: jber@mpi-halle.mpg.de

Jürgen Kirschner
Max-Planck-Intitut für Mikrostrukturphysik
Halle, Germany
e-mail: sekrki@mpi-halle.mpg.de

Coverpicture
see H. Winter, Coincident studies on electronic
interaction mechanisms during scattering of fast
atoms from a LiF(001) surface

This book was carefully produced. Nevertheless,
editors, authors and publisher do not warrant the
information contained therein to be free of errors.
Readers are advised to keep in mind that statements, data, illustrations, procedural details or
other items may inadvertently be inaccurate.

Library of Congress Card No.: applied for
British Library Cataloging-in-Publication Data:
A catalogue record for this book is available from
the British Library

Bibliographic information published by
Die Deutsche Bibliothek
Die Deutsche Bibliothek lists this publication in
the Deutsche Nationalbibliografie; detailed bibli-
ographic data is available in the
Internet at <http://dnb.ddb.de>.

© 2004 WILEY-VCH Verlag GmbH & Co. KGaA,
Weinheim

Printed in the Federal Republic of Germany
Printed on acid-free paper

Composition Uwe Krieg, Berlin
Printing Strauss GmbH, Mörlenbach
Bookbinding Litges & Dopf Buchbinderei
GmbH, Heppenheim

ISBN 3-527-40477-5

ac

Contents

Preface

The understanding of the behavior of interacting many-body systems is a goal with a long history dating back to the beginning of modern physics. The interest in this topic is not only due to the intellectual challenge in describing the complex and the fascinating facets of the many-body dynamics. It is also established by now that various important material properties are driven by the correlated behavior of the system's constituents, as documented by a variety of examples in this book.

The rapid increase in computational resources has triggered an intensive research on the interlink between many-body idealized models and ab-initio single particle approaches. The aim of these theoretical efforts is the development of a realistic theoretical framework capable of capturing many-body effects in electronic materials. This book contains a snapshot on the current status of some of the developments in this direction, namely the implementation of the dynamical mean-field theory and the so-called GW method as well as the calculations based on many-body perturbation theory and the embedded-cluster approach.

On the other hand, in recent years there has been a major advance in the experimental techniques that are capable of imaging the various manifestations of electronic correlation. The most notable progress in this direction is the realization of the coincidence detection of many particles. These experiments allow a detailed insight into the various aspects of the correlated behavior of electronic matter. An overview of the most recent experimental developments is provided by several contributions in this book. A particular emphasis is put on presenting and comparing some of the most prominent techniques presently feasible, namely electron-electron coincidence experiments, electron-ion coincidence techniques, the simultaneous electron-photon detection and photoelectron-Auger electron coincidence spectroscopy.

One of the prime aims of the present book is to highlight the ubiquitous nature of electronic correlations and to stimulate cross fertilization between the various research fields. Therefore, examples are included from atomic, molecular, cluster, surface and solid state physics.

Another aspect which we regard vital for further advances is the close connection between the experimental and the theoretical efforts. Many examples in this book document how theory and experiment can guide each other for finding the appropriate answers and for pointing out to shortcomings and future directions.

The book is based on invited contributions to the conference Coincidence Studies of Surfaces, Thin Films and Nanostructures which took place in September, 2003, at the Ringberg Castle in Germany. We acknowledge with thanks the generous support of the Max-Planck-Institut für Mikrostrukturphysik in Halle and the competent help of H. Schwabe and D. Röder in organizing the workshop.

J. Berakdar and *J. Kirschner*

Halle, 2004

Correlation Spectroscopy of Surfaces, Thin Films, and Nanostructures. Edited by Jamal Berakdar, Jürgen Kirschner
Copyright © 2004 Wiley-VCH Verlag GmbH & Co. KGaA, Weinheim
ISBN: 3-527-40477-5

List of Contributors

- *Oleg Mihailovich Artamonov,* Ch. 6

  Research Institute of Physics,
  St.Petersburg University,
  Uljanovskaja 1, Petrodvoretz,
  198904 St. Petersburg,
  Russia

- *Ferdi Aryasetiawan,* Ch. 1, 7

  Research Institute for Computational Sciences,
  AIST
  Tsukuba Central 2, Umezono 1-1-1
  Tsukuba, Ibaraki 305-8568
  Japan
  e-mail: f-aryasetiawan@aist.go.jp

- *Robert Bartynski,* Ch. 13, 14

  Department of Physics and Astronomy and
  Laboratory of Surface Modification,
  Rutgers University,
  136 Frelinghuysen Road,
  08855 Piscataway, New Jersey
  USA
  e-mail: bart@physics.rutgers.ed

- *Friedhelm Bell,* Ch. 8

  University of Munich,
  Am Coulombwall 1,
  85748 Garching
  Germany
  e-mail: friedhelm.bell@bnw4u.de

- *Jamal Berakdar,* Ch. 3, 5

  Max-Planck Institut für Mikrostrukturphysik
  Weinberg 2,
  06120 Halle
  Germany
  e-mail: jber@mpi-halle.mpg.de

- *Silke Biermann,* Ch. 1

  Centre de Physique Theorique (CPHT)
  Ecole Polytechnique
  91128 Palaiseau Cedex
  France
  e-mail: Silke.Biermann@cpht.polytechnique.fr

- *Joachim Burgdörfer,* Ch. 10

  Institute for Theoretical Physics
  Vienna University of Technology
  Wiedner Hauptstraße 8–10 (E136)
  1040 Vienna
  Austria
  e-mail: burg@concord.itp.tuwien.ac.at

- *Cameron Bowles,* Ch. 7

  Atomic and Molecular Physics Laboratories,
  Research School of Physical Sciences and Engineering,
  Australian National University,
  Canberra, ACT 0200
  Australia

- *C. Creagh,* Ch. 15

  Murdoch University
  2-35 Physical Sciences
  Murdoch
  Australia
  e-mail: thurgate@central.murdoch.edu.au

Correlation Spectroscopy of Surfaces, Thin Films, and Nanostructures. Edited by Jamal Berakdar, Jürgen Kirschner
Copyright © 2004 Wiley-VCH Verlag GmbH & Co. KGaA, Weinheim
ISBN: 3-527-40477-5

- *Michal Dallos,   Ch. 10*

  Institute for Theoretical Chemistry and Structural Biology
  University of Vienna
  Währinger Straße 17/403
  A-1090 Vienna
  Austria
  e-mail: michal.dallos@univie.ac.at

- *Fabiana Da Pieve,   Ch. 14*

  Department of Physics and Unità INFM
  University of Rome, Roma Tre
  Via della Vasca Navale 84
  I-00146 Rome
  Italy
  e-mail: dapieve@fis.uniroma3.it

- *Roland Feder,   Ch. 9*

  Theoretische Festkörperphysik
  Universität Duisburg-Essen,
  D-47048 Duisburg
  Germany
  e-mail: feder3@dagobert.uni-duisburg.de

- *Natasha Fominykh,   Ch. 3*

  Max-Planck Institut für Mikrostrukturphysik
  Weinberg 2,
  06120 Halle
  Germany
  e-mail: fom@mpi-halle.de

- *Antoine Georges,   Ch. 1*

  Laboratoire de Physique Theorique
  de l'Ecole Normale Superieure
  24, rue Lhomond
  75231 Paris Cedex 05
  France
  e-mail: georges@lpt.ens.fr

- *H. Gollisch,   Ch. 9*

  Theoretische Festkörperphysik
  Universität Duisburg-Essen,
  D-47048 Duisburg
  Germany

- *Roberto Gotter,   Ch. 14*

  National Laboratory TASC-INFM
  Area Science Park
  SS 14 Km 163.5
  I-34012 Basovizza, Trieste
  Italy
  e-mail: gotter@tasc.infm.it

- *Steven L. Hulbert,   Ch. 13*

  National Synchrotron Light Sourc
  e Brookhaven National Laboratory
  Upton, New York 11973
  USA
  e-mail: Hulbert@bnl.gov

- *Kouji Isari,   Ch. 16*

  Research Center for Nanodevices and Systems,
  Hiroshima University,
  1-4-2 Kagamiyama, Higashi-Hiroshima 739-8527
  Japan
  e-mail: isari@sxsys.hiroshima-u.ac.jp

- *Stefano Iacobucci,   Ch. 14*

  IMIP-CNR Area della Ricerca di Roma
  via Salaria Km 29,3
  Montelibretti, Rome
  Italy
  e-mail: iacobucci@fis.uniroma3.it

- *Z.-T. Jiang,   Ch. 15*

  Murdoch University
  Perth, Western Australia

- *Anatoli S. Kheifets,   Ch. 7*

  Atomic and Molecular Physics Laboratories,
  Research School of Physical Sciences and Engineering,
  Australian National University,
  Canberra, ACT 0200
  Australia

• *Oleg Kidun,   Ch. 3*

Max-Planck Institut für Mikrostrukturphysik
Weinberg 2,
06120 Halle
Germany
e-mail: okidun@mpi-halle.de

• *Jürgen Kirschner,   Ch. 5*

Max-Planck Institut für Mikrostrukturphysik
Weinberg 2,
06120 Halle
Germany
e-mail: sekrki@mpi-halle.de

• *Eiichi Kobayashi,   Ch. 16*

Institute of Materials Structure Science, KEK,
1-1 Oho, Tsukuba 305-0801
Japan
e-mail: eiichik@post.kek.jp

• *Azzedine Lahmam-Bennani,   Ch. 4*

Laboratoire des Collisions Atomiques et
Moléculaires,
Bâtiment 351
Université de Paris-Sud XI,
F-91405 Orsay cedex
France
e-mail: azzedine.l-bennani@lcam.u-psud.fr

• *Andrea Marini,   Ch. 2*

Departamento de Física de Materiales,
Facultad de Ciencias Químicas,
Universidad del Pais Vasco,

and

Donostia International Physics Center
Universidad del Pais Vasco
E–20018 San Sebastián,
Basque Country
Spain
e-mail: marini@sc.ehu.es

• *Hans Lischka,   Ch. 10*

Institute for Theoretical Chemistry and Struc-
tural Biology
University of Vienna
Währinger Straße 17
A-1090 Vienna
Austria
e-mail: Hans.Lischka@univie.ac.at

• *Andrea Liscio,   Ch. 14*

Department of Physics and Unità INFM
University of Rome, Roma Tre
Via della Vasca Navale 84
I-00146 Rome
Italy
e-mail: liscio@fis.uniroma3.it

and

IMIP-CNR Area della Ricerca di Roma,
via Salaria Km 29,3
Montelibretti
Italy

• *Kazuhiko Mase,   Ch. 16*

Institute of Materials Structure Science, KEK,
1-1 Oho, Tsukuba 305-0801
Japan
e-mail: mase@post.kek.jp

• *Francesco Offi,   Ch. 14*

Department of Physics and Unità INFM
University of Rome, Roma Tre
Via della Vasca Navale 84
I-00146 Rome
Italy
e-mail: offi@fis.uniroma3.it

• *Masahide Ohno,   Ch. 12*

Quantum Science Research,
2-8-5 Tokiwadai,
Itabashi-ku,
Tokyo, 174-0071
Japan

- *G. van Riessen,   Ch. 15*

  Murdoch University
  Perth, Western Australia

- *Alessandro Ruocco,   Ch. 14*

  Department of Physics and Unità INFM
  University of Rome, Roma Tre
  Via della Vasca Navale 84
  I-00146 Rome
  Italy
  e-mail: ruocco@fis.uniroma3.it

- *Sergej Samarin,   Ch. 6*

  School of Physics,
  The University of Western Australia,
  35 Stirling Hwy, Crawley,
  WA 6009,
  Australia
  e-mail: samar@physics.uwa.edu.au

- *Vladimir A. Sashin,   Ch. 7*

  Atomic and Molecular Physics Laboratories,
  Research School of Physical Sciences and Engineering,
  Australian National University,
  Canberra, ACT 0200
  Australia

- *Alex K. See,   Ch. 13*

  Chartered Semiconductor Manufacturing
  Technology Development Division
  Singapore 738406
  Singapore

- *Anthony David Sergeant,   Ch. 6*

  School of Physics,
  The University of Western Australia,
  35 Stirling Hwy, Crawley,
  WA 6009
  Australia

- *Wing-Kit Siu,   Ch. 13*

  Telcordia Technologies
  1 Telcordia Drive
  Somerset, NJ 08854
  USA
  e-mail: wksiu@physics.rutgers.edu

- *Giovanni Stefani,   Ch. 14*

  Department of Physics and Unità INFM
  University of Rome, Roma Tre
  Via della Vasca Navale 84
  I-00146 Rome
  Italy
  e-mail: stefani@fis.uniroma3.it

- *Stephen M. Thurgate,   Ch. 15*

  Murdoch University
  Perth, Western Australia

- *Alberto Verdini,   Ch. 14*

  National Laboratory TASC-INFM
  Area Science Park
  SS 14 Km 163.5
  I-34012 Basovizza, Trieste
  Italy
  e-mail: verdini@tasc.infm.it

- *Maarten Vos,   Ch. 7*

  Atomic and Molecular Physics Laboratories,
  Research School of Physical Sciences and Engineering,
  Australian National University,
  Canberra, ACT 0200
  Australia

- *Erich Weigold,   Ch. 7*

  Atomic and Molecular Physics Laboratories,
  Research School of Physical Sciences and Engineering,
  Australian National University,
  Canberra, ACT 0200
  Australia
  e-mail: erich.weigold@arc.gov.au

- *James Francis Williams,* Ch. 6

  School of Physics,
  The University of Western Australia,
  35 Stirling Hwy, Crawley,
  WA 6009
  Australia
  e-mail: jfw@cyllene.uwa.edu.au

- *Carsten Winkler,* Ch. 5

  Max-Planck Institut für Mikrostrukturphysik
  Weinberg 2,
  06120 Halle
  Germany
  e-mail: winkler@mpi-halle.de

- *Helmut Winter,* Ch. 11

  Institut für Physik der
  Humboldt-Universität zu Berlin,
  Physik der Grenzflächen und dünnen Schichten
  Brook-Taylor-Str. 6
  12489 Berlin, Germany
  e-mail: winter@physik.hu-berlin.de

- *Ludger Wirtz,* Ch. 10

  Donostia International Physics Center (DIPC)
  Paseo Manuel de Lardizabal 4
  20018 San Sebastián
  Spain

  and

  Institute for Theoretical Physics
  Vienna University of Technology
  Wiedner Hauptstraße 8-10/136
  1040 Vienna
  Austria

- *Hua Yao,* Ch. 14

  Department of Physics and Astronomy and
  Laboratory of Surface Modification,
  Rutgers University,
  136 Frelinghuysen Road,
  08855 Piscataway, New Jersey
  e-mail: huayao@physics.rutgers.edu

1 A First-Principles Scheme for Calculating the Electronic Structure of Strongly Correlated Materials: GW+DMFT

Ferdi Aryasetiawan, Silke Biermann, and Antoine Georges

1.1 Introduction

The last few decades have witnessed substantial progress in the field of electronic structure of materials. Using density functional theory (DFT) [1,2] within the local density approximation (LDA) or generalized gradient approximation (GGA) [3] it is quite routine to calculate the electronic structure of relatively complicated materials containing tens of atoms per unit cell. The success of LDA, however, is also accompanied by a number of serious problems. It was noticed very early on that, when applied to calculate the band structures of s–p semiconductors and insulators, the band gaps are systematically underestimated by some tens of percents. Apart from the too small gaps, the band dispersions are very reasonable. This remarkable property of the LDA is still waiting for an explanation since formally there is no theoretical justification for identifying the one-particle Kohn–Sham eigenvalues as quasiparticle energies observed in photoemission experiments. Applications to alkali metals also indicate some problems, albeit less serious. When the band dispersions are compared with photoemission data, they are found to be too wide by 10–30%. Some many-body calculations of the electron gas [4, 5], however, suggest that the band widths are actually widened compared with the free-electron values and that the LDA performs better than is commonly believed. If this turns out to be true, photoemission data would presumably need a complete revision. In any case, the LDA errors in s–p metals are probably less significant than the band gap errors in semiconductors and insulators.

A much more serious problem of the LDA arises when it is applied to calculate the electronic structures of so-called "strongly correlated systems". We have to be more precise with what we mean by correlations. Even in the electron gas, correlation as conventionally defined is rather large. It is as large as exchange so that the two almost cancel each other leaving the free-electron band essentially unchanged. Thus, it is more appropriate in our case to define correlation as anything beyond the LDA rather than anything beyond the Fock exchange since the former is usually our starting point in electronic structure calculations of solids.

Strongly correlated systems are characterized by partially occupied localized orbitals such as found in transition metal oxides or 4f metals. Here the problem is often more of a *qualitative* rather than a *quantitative* nature. It is often found that the LDA predicts a transition metal oxide to be a metal whereas experimentally it is an antiferromagnetic insulator. To cite some examples, $LaMnO_3$, famous for its colossal magnetoresistance, and La_2CuO_4, a well-known parent compound of high-temperature superconductors, are antiferromagnetic insulators but

Correlation Spectroscopy of Surfaces, Thin Films, and Nanostructures. Edited by Jamal Berakdar, Jürgen Kirschner

predicted to be metals by the LDA [6]. In cases where the LDA does predict the correct structure, it is legitimate to ask if the one-particle spectrum is also reproduced correctly. According to the currently accepted interpretation, transition metal oxides may be classified as charge-transfer insulators [7, 8], which are characterized by the presence of occupied and unoccupied 3d bands with the oxygen 2p band in between. The gap is then formed by the oxygen 2p and unoccupied 3d bands, unlike the gap in LDA, which is formed by the 3d states (Mott–Hubbard gap). A more appropriate interpretation is to say that the highest valence state is a charge-transfer state: During photoemission a hole is created in the transition metal site but due to the strong 3d Coulomb repulsion it is energetically more favorable for the hole to hop to the oxygen site despite the cost in energy transfer. A number of experimental data, notably 2p core photoemission resonance, suggest that the charge-transfer picture is more appropriate to describe the electronic structure of transition metal oxides. And of course in the case of 4f metals, the LDA, being a one-particle theory, is totally incapable of yielding the incoherent part of the spectral function or satellite structures.

The above difficulties encountered by the LDA have prompted a number of attempts at improving the LDA. Notable among these is the GW approximation (GWA), developed systematically by Hedin in the early sixties [9]. He showed that the self-energy can be formally expanded in powers of the screened interaction W, the lowest term being iGW, where G is the Green function. Due to computational difficulties, for a long time the applications of the GWA were restricted to the electron gas but with the rapid progress in computer power, applications to realistic materials eventually became possible about two decades ago. Numerous applications to semiconductors and insulators reveal that in most cases the GWA [10, 11] removes a large fraction of the LDA band-gap error. Applications to alkali metals show band narrowing from the LDA values and account for more than half of the LDA error (although controversy about this issue still remains [12]).

The success of the GWA in sp materials has prompted further applications to more strongly correlated systems. For this type of materials the GWA has been found to be less successful. For example, GW calculation on nickel [13] does reproduce the photoemission quasiparticle band structure rather well, as compared with the LDA one where the 3d band width is too large by about 1 eV, but the too large LDA exchange splitting of 0.6 eV (experimentally 0.3 eV) remains essentially unchanged. Moreover, the famous 6 eV satellite is not reproduced. Application to NiO [14], a prototype of transition metal oxides, also reveals some shortcomings. One problem is related to the starting Green's function, usually constructed from the LDA Kohn–Sham orbitals and energies. In the LDA the band gap is very small, about 0.2 eV compared with the 4 eV experimental band gap. A commonly used procedure of performing a one-iteration GW calculation yields about 1 eV gap, much too small. This problem is solved by performing a partial self-consistency, where knowledge of the self-energy from the previous iteration is used to construct a better starting one-particle Hamiltonian [14]. This procedure improves the band gap considerably to a self-consistent value of 5.5 eV and at the same time increases the LDA magnetic moment from 0.9 μ_B to about 1.6 μ_B much closer to the experimental value of 1.8 μ_B . However, the GWA maintains the Mott–Hubbard gap, i.e., the gap is formed by the 3d states as in the LDA, instead of the charge-transfer gap. In other words, the top of the valence band is dominated by the Ni 3d. A more recent calculation using a more refined procedure of partial self-consistency has also confirmed these results [15]. The problem with the GWA appears to arise from inadequate account of short-range correlations, probably not properly treated in the random-phase approximation (RPA).

Attempts at improving the LDA to treat strongly correlated systems were initiated by the LDA+U method [16–19], which introduces, on top of the LDA Hamiltonian, a Hubbard U term and a double-counting correction term, usually applied to partially filled 3d or 4f shells. The LDA+U method is essentially a Hartree–Fock approximation to the LDA+U Hamiltonian. In the LDA, the Kohn–Sham potential does not explicitly distinguish between occupied and unoccupied orbitals so that they experience the same potential. In, for example, transition metal oxides, where the 3d orbitals are partially occupied, this leads to metallicity or to underestimation of the band gap. The LDA+U cures this problem by approximately pushing down the occupied orbitals by $U/2$ and pushing up the unoccupied orbitals by $U/2$, creating a lower and upper Hubbard band, thus opening up a gap of the order of the Hubbard U. The LDA+U method has been successfully applied to late transition metal oxides, rare earth compounds such as CeSb, as well as to problems involving metal–insulator transition and charge-orbital ordering.

More recently, the idea of the LDA+U was extended further by treating the Hubbard U term in a more sophisticated fashion utilizing the dynamical mean-field theory (DMFT) [20]. The DMFT is remarkably well suited for treating systems with strong on-site correlations because the on-site electronic Coulomb interactions are summed to all orders. This is achieved by using a mapping onto a self-consistent quantum impurity problem, thereby including the effects of the surrounding in a mean-field approximation. The strength of the DMFT is its ability to properly describe Mott phenomenon or the formation of local moments, which is the key to understanding many physical properties in strongly correlated materials. The combination of LDA and DMFT takes advantage of the first-principles nature of LDA while at the same time incorporates local correlation effects not properly treated within the LDA. The LDA+DMFT method [19, 21, 22] has now been successfully applied to a number of systems.

In both the LDA+U and LDA+DMFT methods, two fundamental problems remain unaddressed. First, the Hubbard U is usually treated as a parameter, and second, the Hubbard U term contains interaction already included in the LDA but it is not clear how to take into account this double-counting term in a precise way. Thus, a truly first-principles theory for treating strongly correlated systems is still lacking. In this chapter, we describe a dynamical mean-field approach for calculating the electronic structure of strongly correlated materials from first-principles [23, 24]. The DMFT is combined with the GW method, which enables one to treat strong interaction effects [25]. One of the main features of the new scheme is that the Hubbard U is calculated from first principles through a self-consistency requirement on the on-site screened Coulomb interaction, analogous to the self-consistency in the local Green's function in the DMFT. Since the GWA has an explicit diagrammatic representation, the on-site contribution of the GW self-energy can be readily identified and the scheme then allows for a precise double-counting correction.

In the next two sections, we will give a summary of the GWA and DMFT, describing their main features. In the fourth section we lay out the GW+DMFT scheme, followed by a simplified application of the scheme to the excitation spectrum of nickel. Finally we discuss some future challenges and directions.

1.2   The *GW* Approximation

1.2.1   Theory

It can be shown that the self-energy may be expressed as [9]

$$\Sigma(1,2) = -i \int d3\, d4\, v(1,4) G(1,3) \frac{\delta G^{-1}(3,2)}{\delta\phi(4)} \tag{1.1}$$

where v is the bare Coulomb interaction, G is the Green function and ϕ is an external time-dependent probing field. We have used the short-hand notation $1 = (x_1 t_1)$. From the equation of motion of the Green function

$$G^{-1} = i\frac{\partial}{\partial t} - H_0 - \Sigma \tag{1.2}$$

$$H_0 = h_0 + \phi + V_{\rm H} \tag{1.3}$$

h_0 is the kinetic energy and $V_{\rm H}$ is the Hartree potential. We then obtain

$$\begin{aligned}
\frac{\delta G^{-1}(3,2)}{\delta\phi(4)} &= -\delta(3-2)\left[\delta(3-4) + \frac{\delta V_{\rm H}(3)}{\delta\phi(4)}\right] - \frac{\delta\Sigma(3,2)}{\delta\phi(4)} \\
&= -\delta(3-2)\epsilon^{-1}(3,4) - \frac{\delta\Sigma(3,2)}{\delta\phi(4)}
\end{aligned} \tag{1.4}$$

where ϵ^1 is the inverse dielectric matrix. The GWA is obtained by neglecting the vertex correction $\delta\Sigma/\delta\phi$, which is the last term in Eq. (1.4). This is just the random-phase approximation (RPA) for ϵ^{-1}. This leads to

$$\Sigma(1,2) = iG(1,2)W(1,2) \tag{1.5}$$

where we have defined the screened Coulomb interaction W by

$$W(1,2) = \int d3 v(1,3)\epsilon^{-1}(3,2) \tag{1.6}$$

The RPA dielectric function is given by

$$\epsilon = 1 - vP \tag{1.7}$$

where

$$\begin{aligned}
P(\boldsymbol{r},\boldsymbol{r}';\omega) &= -2i \int \frac{d\omega'}{2\pi} G(\boldsymbol{r},\boldsymbol{r}';\omega+\omega')G(\boldsymbol{r}',\boldsymbol{r};\omega') \\
&= 2\sum_i^{occ}\sum_j^{unocc} \psi_i(\boldsymbol{r})\psi_i^*(\boldsymbol{r}')\psi_j^*(\boldsymbol{r})\psi_j(\boldsymbol{r}') \\
&\quad \times \left\{\frac{1}{\omega - \varepsilon_j + \varepsilon_i + i\delta} - \frac{1}{\omega + \varepsilon_j - \varepsilon_i - i\delta}\right\}
\end{aligned} \tag{1.8}$$

with the Green function constructed from a one-particle band structure $\{\psi_i, \varepsilon_i\}$. The factor of 2 arises from the sum over spin variables. In frequency space, the self-energy in the GWA takes the form

$$\Sigma(r, r'; \omega) = \frac{i}{2\pi} \int d\omega' e^{i\eta\omega'} G(\boldsymbol{r}, \boldsymbol{r}'; \omega + \omega') W(\boldsymbol{r}, \boldsymbol{r}'; \omega') \tag{1.9}$$

We have so far described the zero temperature formalism. For finite temperature we have

$$P(\boldsymbol{r}, \boldsymbol{r}'; i\nu_n) = \frac{2}{\beta} \sum_{\omega_k} G(\boldsymbol{r}, \boldsymbol{r}'; i\nu_n + i\omega_k) G(\boldsymbol{r}', \boldsymbol{r}; i\omega_k) \tag{1.10}$$

$$\Sigma(r, r'; i\omega_n) = -\frac{1}{\beta} \sum_{\nu_k} G(r, r'; i\omega_n + i\nu_k) W(r, r'; i\nu_k) \tag{1.11}$$

In the Green function language, the Fock exchange operator in the Hartree–Fock approximation (HFA) can be written as iGv. We may therefore regard the GWA as a generalization of the HFA, where the bare Coulomb interaction v is replaced by a screened interaction W. We may also think of the GWA as a mapping to a polaron problem where the electrons are coupled to some bosonic excitations (e.g., plasmons) and the parameters in this model are obtained from first-principles calculations.

The replacement of v by W is an important step in solids where screening effects are generally rather large relative to exchange, especially in metals. For example, in the electron gas, within the GWA, exchange and correlation are approximately equal in magnitude, to a large extent canceling each other, modifying the free-electron dispersion slightly. But also in molecules, accurate calculations of the excitation spectrum cannot neglect the effects of correlations or screening. The GWA is physically sound because it is qualitatively correct in some limiting cases [26].

1.2.2 The GW Approximation in Practice

The quality of the GWA may be seen in Figure 1.1, where a plot of band gaps of a number of well known semiconductors and insulators is displayed. It is clear from the plot that the LDA systematically underestimates the band gaps and that the GWA substantially improves the LDA band gaps. It has been found that for some materials, like MgO and InN, significant error still remains within the GWA. The reason for the discrepancy has not been understood well. One possible explanation is that the result of the one-iteration GW calculation may depend on the starting one-particle band structure. For example, in the case of InN, the starting LDA band structure has no gap. This may produce a metal-like (over)screened interaction W which fails to open up a gap or yields too small a gap in the GW calculation. Similar behavior is also found in the more extreme case of NiO, where a one-iteration GW calculation only yields a gap of about 1 eV starting from an LDA gap of 0.2 eV (the experimental gap is 4 eV) [10, 14].

The problems with the GWA arise when it is applied to strongly correlated systems. Application to ferromagnetic nickel [13] illustrates some of the difficulties with the GWA. Starting from the LDA band structure, a one-iteration GW calculation does improve significantly the LDA band structure. In particular it reduces the too large 3d band width bringing it into

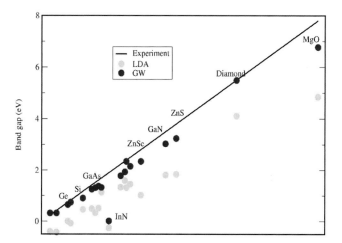

Figure 1.1: Band gaps of some selected semiconductors and insulators calculated within the GWA compared with the LDA and experimental values. The GW data are taken from [27].

much better agreement with photoemission data. However, the too large LDA exchange splitting (0.6 eV compared with the experimental value of 0.3 eV) remains essentially unchanged. Moreover, the famous 6 eV satellite, which is of course missing in the LDA, is not reproduced. These problems point to deficiencies in the GWA in describing short-range correlations since we expect that both exchange splitting and satellite structure are influenced by on-site interactions. In the case of exchange splitting, long-range screening also plays a role in reducing the HF value and the problem with the exchange splitting indicates a lack of spin-dependent interaction in the GWA: In the GWA the spin dependence only enters in G not in W.

Application to NiO, the prototype of Mott–Hubbard transition metal oxides, reveals another difficulty with the one-iteration GWA. As already mentioned previously, when the starting band structure is far from the experimental quasiparticle band structure, a one-iteration GW calculation may not be sufficient. This problem may be circumvented by performing a partial self-consistent calculation in which the self-energy from the previous iteration at a given energy, such as the Fermi energy of the center of the band of interest, is used to construct a new set of one-particle orbitals. This procedure is continued to self-consistency such that the starting one-particle band structure gives zero self-energy correction [10, 14, 15]. A more serious problem, however, is describing the charge-transfer character of the top of the valence band. The GWA essentially still maintains the Mott–Hubbard band gap as in the LDA, i.e., the top of the valence band is mainly of 3d character rather than the charge-transfer character dominated by the 2p oxygen hole. As in nickel, the problem with the satellite arises again. Depending on the starting band structure, a satellite may be reproduced albeit at a too high energy. Thus there is a strong need to improve the short-range correlations in the GWA which may be achieved by using a suitable approach based on the dynamical mean-field theory described in the next section.

1.3   Dynamical Mean Field Theory

Dynamical mean field theory (DMFT) [20] was originally developed within the context of *models* for correlated fermions on a lattice, where it has proven very successful for determining the phase diagrams or for calculations of excited states properties. It is a non-perturbative method and as such appropriate for systems with any strength of the interaction. In recent years, combinations of DMFT with band structure theory, in particular density functional theory with the local density approximation (LDA) have emerged. The idea is to correct for shortcomings of DFT-LDA due to strong Coulomb interactions and localization (or partial localization) phenomena that cause effects very different from a homogeneous itinerant behavior. Such signatures of correlations are well-known in transition metal oxides or f-electron systems but are also present in several elemental transition metals.

The application of DMFT to *real solids* relies on a *representability conjecture* assuming that local quantities, for example the local Green's function or self-energy of a solid, can be calculated from a local impurity model, that is one can find a *dynamical mean field* $\mathcal{G}_0$ and a Hubbard parameter U, such that the Green's function calculated from the effective action

$$
\begin{aligned}
S = {} & \int_0^\beta \mathrm{d}\tau \sum_{m\sigma} c_{m\sigma}^\dagger(\tau) \mathcal{G}_{0mm'\sigma}^{-1}(\tau-\tau') c_{m'\sigma}(\tau') \\
& + \frac{1}{2} \int_0^\beta \mathrm{d}\tau \sum_{mm'\sigma} U_{mm'} n_{m\sigma}(\tau) n_{m'-\sigma}(\tau) \\
& + \frac{1}{2} \int_0^\beta \mathrm{d}\tau \sum_{m\neq m'\sigma} (U_{mm'} - J_{mm'}) n_{m\sigma}(\tau) n_{m'\sigma}(\tau)
\end{aligned}
\tag{1.12}
$$

coincides with the local Green's function of the solid. For a model of correlated fermions on a lattice with infinite coordination number this conjecture can rigorously be proven: it is a consequence of the absence of any non-local contributions to the self-energy of the system. For a real solid the situation is somewhat more complicated, not only due to the finite coordination number but also to the difficulty in defining the notion of locality. This notion is crucial at both stages, for the construction of the impurity model, where *long-range* Coulomb interactions are mimicked by *local* Hubbard parameters[1], and for the resolution of the model within DMFT, which approximates the full self-energy of the model by a local quantity. Applications of DMFT to electronic structure calculations (e.g. the LDA+DMFT method) are therefore always defined within a specific basis set using localized basis functions. Within an LMTO implementation for example locality can naturally be defined as referring to the same muffin tin sphere. This amounts to defining matrix elements $G_{L\boldsymbol{R},L'\boldsymbol{R}'}(i\omega)$ of the full Green's function

$$
G(\boldsymbol{r},\boldsymbol{r}',i\omega) = \sum_{LL'\boldsymbol{R}\boldsymbol{R}'} \chi_{L\boldsymbol{R}}^*(\boldsymbol{r}) G_{L\boldsymbol{R},L'\boldsymbol{R}'}(i\omega) \chi_{L'\boldsymbol{R}'}(\boldsymbol{r}')
$$

and assuming that its local, that is "on-sphere" part equals the Green's function of the local impurity model (1.12). Here $\boldsymbol{R}, \boldsymbol{R}'$ denote the coordinates of the centers of the muffin tin

---

[1] For a discussion of the appropriateness of local Hubbard parameters see [28].

spheres, while r, r' can take any values. The index $L = (n, l, m)$ regroups all radial and angular quantum numbers. The dynamical mean field $\mathcal{G}_0$ in Eq. (1.12) has to be determined in such a way that the Green's function $G_{\text{impurity} L, L', \sigma}$ of the impurity model Eq. (1.12) coincides with $G_{L\boldsymbol{R}, L'\boldsymbol{R}'}(i\omega)$ if the impurity model self-energy is used as an estimate for the true self-energy of the solid. This self-consistency condition reads

$$G_{\text{impurity}}(i\omega_n) = \sum_k \left(i\omega_n + \mu - H_o(k) - \Sigma(i\omega_n)\right)^{-1}$$

where Σ, H_0 and G are matrices in orbital and spin space, and $i\omega + \mu$ is a matrix proportional to the unit matrix in that space.

Together with Eq. (1.12) this defines the DMFT equations that have to be solved self-consistently. Note that the main approximation of DMFT is hidden in the self-consistency condition where the local self-energy has been promoted to the full lattice self-energy.

The representability assumption can actually be extended to other quantities of a solid than its local Green's function and self-energy. In "extended DMFT" [29–32] e.g. a two particle correlation function is calculated and can then be used in order to represent the local screened Coulomb interaction W of the solid. This is the starting point of the "GW+DMFT" scheme described in Section 1.4.

1.3.1 DMFT in Practice

Combinations of DFT-LDA and DMFT, so-called "LDA+DMFT" techniques [21], have so far been applied to transition metals (Fe, Ni, Mn) and their oxides (e.g. La/YTiO$_3$, V$_2$O$_3$, Sr/CaVO$_3$) as well as elemental f-electron materials (Pu, Ce) and their compounds. In the most general formulation, one starts from a many-body Hamiltonian of the form

$$\begin{aligned}
H = &\sum_{\{im\sigma\}} (H^{\text{LDA}}_{im,i'm'} - H^{\text{dc}})a^{+}_{im\sigma}a_{i'm'\sigma} \\
&+ \frac{1}{2}\sum_{imm'\sigma} U^{i}_{mm'}n_{im\sigma}n_{im'-\sigma} \\
&+ \frac{1}{2}\sum_{im\neq m'\sigma} (U^{i}_{mm'} - J^{i}_{mm'})n_{im\sigma}n_{im'\sigma},
\end{aligned}$$

(1.13)

where H^{LDA} is the effective Kohn–Sham Hamiltonian derived from a self-consistent DFT-LDA calculation. This one-particle Hamiltonian is then corrected by Hubbard terms for direct and exchange interactions for the "correlated" orbitals, e.g. d or f orbitals. In order to avoid double counting of the Coulomb interactions for these orbitals, a correction term H^{dc} is subtracted from the original LDA Hamiltonian. The resulting Hamiltonian (1.13) is then treated within dynamical mean field theory by assuming that the many-body self-energy associated with the Hubbard interaction terms can be calculated from a multi-band impurity model.

This general scheme can be simplified in specific cases, e.g. in systems with a separation of the correlated bands from the "uncorrelated" ones, an effective model of the correlated bands can be constructed; symmetries of the crystal structure can be used to reduce the number of components of the self-energy etc.

Despite the huge progress made in the understanding of the electronic structure of correlated materials thanks to such LDA+DMFT schemes, certain conceptual problems remain open: These are related to the choice of the Hubbard interaction parameters and to the double counting corrections. An a priori choice of which orbitals are treated as correlated and which orbitals are left uncorrelated has to be made, and the values of U and J have to be fixed. Attempts at calculating these parameters from constrained LDA techniques are appealing in the sense that one can avoid introducing external parameters to the theory, but suffer from the conceptual drawback in that screening is taken into account in a static manner only [28]. Finally, the double counting terms are necessarily ill defined due to the impossibility of singling out in the LDA treatment contributions to the interactions stemming from specific orbitals. These drawbacks of LDA+DMFT provide a strong motivation to attempt the construction of an electronic structure method for correlated materials beyond combinations of LDA and DMFT.

1.4   GW+DMFT

The idea behind combining the GWA and the DMFT is to take advantage of the strong features of the two theories. The GWA, being based on RPA screening, is capable of taking into account long-range correlations but does not describe properly short-range correlations. On the other hand, the strength of the DMFT is its ability to describe on-site correlations by means of a mapping of the many-body problem onto an impurity problem where the on-site interactions are summed to all orders. While the GWA is a fully first-principles theory, the DMFT has been traditionally used in conjunction with a Hubbard model. Clearly, it is desirable to combine the two theories into a consistent theory where the parameters in the Hubbard model are determined self-consistently from first-principles via the GWA.

Recently, first steps have been undertaken towards a combination of the GWA and DMFT, both in a model context [24] and within the framework of realistic electronic structure calculations [23]. The basic physical idea of GW+DMFT is to separate the lattice into an on-site part and the rest. The on-site self-energy is taken to be the impurity self-energy calculated by the DMFT and the off-site self-energy is calculated by the GWA. Viewed from the GWA, we replace the on-site GW self-energy by that of DMFT, correcting the GW treatment of on-site correlations. Viewed from the DMFT, we add off-site contributions to the self-energy approximated within the GWA, giving a momentum dependent self-energy.

The impurity problem contains a Hubbard interaction U that is usually treated as a parameter. In order to calculate the U, we introduce a self-consistency condition that the U screened by the effective bath in the impurity model be equal to the local projection of the global screened interaction W. This condition is complementary to the condition imposed in the DMFT that the impurity Green's function be equal to the local Green's function.

The above physical ideas can be formulated in a precise way using the free-energy functional of Luttinger and Ward (LW). A generalization of the original LW functional takes the form [33] (see also Ref. [34])

$$\Gamma(G, W) = \mathrm{Tr} \ln G - \mathrm{Tr}[(G_{\mathrm{H}}^{-1} - G^{-1})G] - \frac{1}{2} \mathrm{Tr} \ln W$$
$$+ \frac{1}{2} \mathrm{Tr}[(v^{-1} - W^{-1})W] + \Psi[G, W] \tag{1.14}$$

$G_{\mathrm{H}}^{-1} = i\omega_n + \mu + \nabla^2/2 - V_{\mathrm{H}}$ corresponds to the Hartree Green's function with V_{H} being the Hartree potential. The functional $\Psi[G, W]$ is a generalization of the original LW $\Phi[G]$ functional, whose derivative with respect to G gives the self-energy. A more general derivation of (1.14) using a Hubbard-Stratonovich transformation and a Legendre transformation with respect to both G and W may be found in a later work [34]. It is straightforward to verify that at equilibrium the stationarity of Γ yields

$$\frac{\delta\Gamma}{\delta G} = 0 \rightarrow G^{-1} = G_{\mathrm{H}}^{-1} - \Sigma, \qquad \Sigma = \frac{\delta\Psi}{\delta G} \tag{1.15}$$

$$\frac{\delta\Gamma}{\delta W} = 0 \rightarrow W^{-1} = v^{-1} - P, \qquad P = -2\frac{\delta\Psi}{\delta W} \tag{1.16}$$

The functional Ψ is divided into on-site and off-site components:

$$\Psi = \Psi_{GW}^{\text{off-site}}\left[G^{RR'}, W^{RR'}\right] + \Psi_{\text{imp}}^{\text{on-site}}\left[G^{RR}, W^{RR}\right] \tag{1.17}$$

where R denotes a lattice site. In the GWA the functional Ψ is given by

$$\Psi_{GW}[G, W] = \frac{1}{2}GWG \tag{1.18}$$

The impurity part Ψ_{imp} is generated from a local quantum impurity problem defined on a single atomic site with an effective action

$$S = \int \mathrm{d}\tau \mathrm{d}\tau' \left[-\sum c_L^+(\tau) \mathcal{G}_{LL'}^{-1}(\tau - \tau') c_{L'}(\tau') \right.$$
$$\left. + \frac{1}{2} \sum : c_{L_1}^+(\tau)c_{L_2}(\tau) : \mathcal{U}_{L_1 L_2 L_3 L_4}(\tau - \tau') : c_{L_3}^+(\tau')c_{L_4}(\tau') : \right] \tag{1.19}$$

The double dots denote normal ordering and L refers to an orbital of angular momentum L on a given sphere where the impurity problem is defined. These orbitals are usually partially filled localized 3d or 4f orbitals.

The GW+DMFT set of equations can now be readily derived from Eqs. (1.14), (1.17), and (1.18) by taking functional derivatives of Ψ with respect to G and W as in Eqs. (1.15) and (1.16):

$$\Sigma = \Sigma_{GW}^{RR'}(1 - \delta_{RR'}) + \Sigma_{\text{imp}}^{RR}\delta_{RR'} \tag{1.20}$$

$$P = P_{GW}^{RR'}(1 - \delta_{RR'}) + P_{\text{imp}}^{RR}\delta_{RR'} \tag{1.21}$$

In practice the self-energy is expanded in some basis set $\{\phi_L\}$ localized in a site. The polarization function on the other hand is expanded in a set of two-particle basis functions $\{\phi_L\phi_{L'}\}$

(product basis) since the polarization corresponds to a two-particle propagator. For example, when using the linear muffin-tin orbital (LMTO) band-structure method, the product basis consists of products of LMTOs. These product functions are generally linearly dependent and a new set of optimized product basis (OPB) [10] is constructed by forming linear combinations of product functions, eliminating the linear dependences. We denote the OPB set by $B_\alpha = \sum_{LL'} \phi_L \phi_{L'} c_{LL'}^\alpha$. To summarize, one-particle quantities like G and Σ are expanded in $\{\phi_L\}$ whereas two-particle quantities such as P and W are expanded in the OPB set $\{B_\alpha\}$. It is important to note that the number of $\{B_\alpha\}$ is generally smaller than the number of $\{\phi_L \phi_{L'}\}$ so that quantities expressed in $\{B_\alpha\}$ can be expressed in $\{\phi_L \phi_{L'}\}$, but not vice versa. In momentum space, Eqs. (1.20) and (1.21) read

$$\Sigma^{LL'}(\boldsymbol{k}, i\omega_n) = \Sigma_{GW}^{LL'}(\boldsymbol{k}, i\omega_n) - \sum_{\boldsymbol{k}} \Sigma_{GW}^{LL'}(\boldsymbol{k}, i\omega_n) + \Sigma_{\text{imp}}^{LL'}(i\omega_n) \tag{1.22}$$

$$P^{\alpha\beta}(\boldsymbol{k}, i\omega_n) = P_{GW}^{\alpha\beta}(\boldsymbol{k}, i\omega_n) - \sum_{\boldsymbol{k}} P_{GW}^{\alpha\beta}(\boldsymbol{k}, i\omega_n) + P_{\text{imp}}^{\alpha\beta}(i\omega_n) \tag{1.23}$$

The second terms in the above two equations remove the on-site contributions of the GW self-energy and polarization, which are already included in Σ_{imp} and P_{imp}.

We are now in a position to outline the self-consistency loop which determines $\mathcal{G}$ and $\mathcal{U}$ as well as the full G and W self-consistently.

- The impurity problem (1.19) is solved, for a given choice of Weiss field $\mathcal{G}_{LL'}$ and Hubbard interaction $\mathcal{U}_{\alpha\beta}$: the "impurity" Green's function

$$G_{\text{imp}}^{LL'} \equiv -\langle T_\tau c_L(\tau) c_{L'}^+(\tau')\rangle_S \tag{1.24}$$

is calculated, together with the impurity self-energy

$$\Sigma_{\text{imp}} \equiv \delta\Psi_{\text{imp}}/\delta G_{\text{imp}} = \mathcal{G}^{-1} - G_{\text{imp}}^{-1}. \tag{1.25}$$

The two-particle correlation function

$$\chi_{L_1 L_2 L_3 L_4} = \langle : c_{L_1}^\dagger(\tau) c_{L_2}(\tau) :: c_{L_3}^\dagger(\tau') c_{L_4}(\tau') :\rangle_S \tag{1.26}$$

must also be evaluated.

- The impurity effective interaction is constructed as follows:

$$W_{\text{imp}}^{\alpha\beta} = \mathcal{U}_{\alpha\beta} - \sum_{L_1 \cdots L_4} \sum_{\gamma\delta} \mathcal{U}_{\alpha\gamma} O_{L_1 L_2}^\gamma \chi_{L_1 L_2 L_3 L_4} [O_{L_3 L_4}^\delta]^* \mathcal{U}_{\delta\beta} \tag{1.27}$$

Here all quantities are evaluated at the same frequency[2] and $O_{L_1 L_2}^\gamma$ is the overlap matrix $\langle B_\gamma | \phi_L \phi_{L'} \rangle$. The polarization operator of the impurity problem is then obtained as:

$$P_{\text{imp}} \equiv -2\delta\Psi_{\text{imp}}/\delta W_{\text{imp}} = \mathcal{U}^{-1} - W_{\text{imp}}^{-1}, \tag{1.28}$$

where the matrix inversions are performed in the OPB $\{B_\alpha\}$.

---

[2] Note that $\chi_{L_1 \ldots L_4}$ does *not* denote the matrix element $< L_1 L_2 | \chi | L_3 L_4 >$, but is rather defined by $\chi(r, r') = \sum_{L_1 .. L_4} \phi_{L_1}^*(r) \phi_{L_2}^*(r) \chi_{L_1 \ldots L_4} \phi_{L_3}(r') \phi_{L_4}(r')$.

- From Eqs. (1.22) and (1.23) the full $\boldsymbol{k}$-dependent Green's function $G(\boldsymbol{k}, i\omega_n)$ and effective interaction $W(\mathbf{q}, i\nu_n)$ can be constructed. The self-consistency condition is obtained, as in the usual DMFT context, by requiring that the on-site components of these quantities coincide with G_{imp} and W_{imp}. In practice, this is done by computing the on-site quantities

$$G_{\text{loc}}(i\omega_n) = \sum_{\boldsymbol{k}}[G_{\text{H}}^{-1}(\boldsymbol{k}, i\omega_n) - \Sigma(\boldsymbol{k}, i\omega_n)]^{-1} \tag{1.29}$$

$$W_{\text{loc}}(i\nu_n) = \sum_{\boldsymbol{q}}[V_{\boldsymbol{q}}^{-1} - P(\boldsymbol{q}, i\nu_n)]^{-1} \tag{1.30}$$

and using them to update the Weiss dynamical mean field $\mathcal{G}$ and the impurity model interaction $\mathcal{U}$ according to:

$$\mathcal{G}^{-1} = G_{\text{loc}}^{-1} + \Sigma_{\text{imp}} \tag{1.31}$$

$$\mathcal{U}^{-1} = W_{\text{loc}}^{-1} + P_{\text{imp}} \tag{1.32}$$

This cycle is iterated until self-consistency for $\mathcal{G}$ and $\mathcal{U}$ is obtained (as well as on G, W, Σ^{xc} and P). When self-consistency is reached, $G_{\text{imp}} = G_{\text{loc}}$ and $W_{\text{imp}} = W_{\text{loc}}$. This implies that at self-consistency, the second term in Eq. (1.22) can be rewritten as (in imaginary-time)

$$\sum_{\boldsymbol{k}} \Sigma_{GW}^{LL'}(\boldsymbol{k}, \tau) = -\sum_{L_1 L_1'} W_{\text{imp}}^{LL_1L'L_1'}(\tau) G_{\text{imp}}^{L_1'L_1}(\tau) \tag{1.33}$$

This shows that the local or on-site contribution of the GW self-energy is precisely subtracted out, thus avoiding double counting. Eventually, self-consistency over the local electronic density can also be implemented, (in a similar way as in LDA+DMFT [35, 36]) by recalculating $\rho(\boldsymbol{r})$ from the Green's function at the end of the convergence cycle above, and constructing an updated Hartree potential. This new density is used as an input of a new GW calculation, and convergence over this external loop must be reached. While implementing self-consistency within the GWA is known to yield unsatisfactory spectra [37], we expect a more favorable situation in the proposed GW+DMFT scheme since part of the interaction effects are treated to all orders.

1.4.1   Simplified Implementation of GW+DMFT and Application to Ferromagnetic Nickel

The full implementation of the proposed approach in a fully dynamical and self-consistent manner is at the present stage computationally very demanding and we regard it as a major challenge for future research. Here, we apply a simplified scheme of the approach [23] to the electronic structure of nickel in order to demonstrate its feasibility and potential. The main simplifications made are:

1. The DMFT local treatment is applied only to the d-orbitals, and we replace the dynamical impurity problem by its static limit, solving the impurity model (1.19) for a frequency-independent $\mathcal{U} = \mathcal{U}(\omega = 0)$.

2. We perform a one-iteration GW calculation in the form [10]: $\Sigma_{GW} = G_{\mathrm{LDA}} \cdot W[G_{\mathrm{LDA}}]$, from which the off-site part of the self-energy is obtained.

The local Green's function is taken to be

$$G_{\mathrm{loc}}^{\sigma}(i\omega_n) = \sum_{\boldsymbol{k}} \left[G_{\mathrm{H}}^{-1}(\boldsymbol{k}, i\omega_n) - \Sigma_{GW}^{\mathrm{off\text{-}site}} \right. \qquad (1.34)$$
$$\left. - \left(\Sigma_{\mathrm{imp},\sigma} - \frac{1}{2}\mathrm{Tr}_{\sigma}\, \Sigma_{\mathrm{imp},\sigma}(0) + V_{xc}^{\mathrm{on\text{-}site}} \right) \right]^{-1}$$

Thus, the off-site part is obtained from the GW self-energy whereas the on-site part is derived from the impurity self-energy with a double-counting correction of the form proposed in [38].

We have performed finite temperature GW and LDA+DMFT calculations (within the LMTO-ASA [39] with 29 irreducible $\boldsymbol{k}$-points) for ferromagnetic nickel (lattice constant 6.654 a.u.), using 4s4p3d4f states, at the Matsubara frequencies $i\omega_n$ corresponding to $T = 630$ K, just below the Curie temperature. The GW self-energy is calculated from a paramagnetic Green's function, leaving the spin-dependence to the impurity self-energy. The resulting self-energies are inserted into Eq. (1.34), which is then used to calculate a new Weiss field according to Eq. (1.31). The Green's function $G_{\mathrm{loc}}^{\sigma}(\tau)$ is recalculated from the impurity effective action by QMC and analytically continued using the maximum entropy algorithm. The resulting spectral function is plotted in Figure (1.2). Comparison with the LDA+DMFT results in [38] shows that the good description of the satellite structure, exchange splitting and band narrowing is indeed retained within the (simplified) GW+DMFT scheme. We have also calculated the quasiparticle band structure, from the poles of Eq. (1.34), after linearization of $\Sigma(\boldsymbol{k}, i\omega_n)$ around the Fermi level[3]. Figure (1.3) shows a comparison of GW+DMFT with the LDA and experimental band structure. It is seen that GW+DMFT correctly yields the bandwidth reduction compared to the (too large) LDA value and renormalizes the bands in a ($\boldsymbol{k}$-dependent) manner.

Because of the static approximation 3), we could not implement self-consistency on W_{loc} (Eq. (1.30)). We chose the value of $\mathcal{U}(\omega = 0)$ ($\simeq 3.2$ eV) by calculating the correlation function χ and ensuring that Eq. (1.27) is fulfilled at $\omega = 0$, given the GW value for $W_{\mathrm{loc}}(\omega = 0)$ ($\simeq 2.2$ eV for nickel [40]).

1.5 Conclusions

The proposed GW+DMFT scheme avoids the conceptual problems inherent to "LDA+DMFT" methods, such as double counting corrections and the use of Hubbard parameters assigned to correlated orbitals. The notion of a self-consistency condition on the on-site Green's function in the DMFT is extended to the screened interaction. Analogous to the usual condition that the impurity Green's function be equal to the on-site Green's function, we demand that the on-site screened Hubbard $\mathcal{U}$ of the impurity be equal to the on-site projection of the global screened interaction W. In this fashion, the Hubbard $\mathcal{U}$ is determined from first-principles. Since the

---

[3] Note however that this linearization is no longer meaningful at energies far away from the Fermi level. We therefore use the unrenormalized value for the quasi-particle residue for the s-band ($Z_s = 1$).

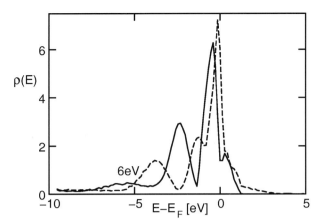

Figure 1.2: Partial density of states of 3d orbitals of nickel (solid/dashed lines give the majority/minority spin contribution) as obtained from the combination of GW and DMFT [23]. For comparison with LDA and LDA+DMFT results, see Ref. [38]; for experimental spectra, see Ref. [42].

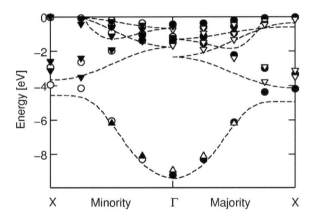

Figure 1.3: band structure of nickel (majority and minority spins) from GW+DMFT scheme [23] (circles) in comparison with the LDA band structure (dashed lines) and experiments [41] (triangles down) and [42] (triangles up).

GWA has a well-defined diagrammatic interpretation, it is also possible to precisely take into account the double-counting correction.

A number of issues are of immediate importance. Solving impurity models with frequency-dependent interaction parameters [24, 43, 44] is one of the most urgent tasks as well as studies of various possible self-consistency schemes. To study these aspects, we are now carrying out GW+DMFT calculations on the electron gas. Applications to real materials are both theoretically and computationally very challenging. Extension to the two-particle Green's function is another field for future research [45].

Acknowledgments

This work was supported in part by NAREGI Nanoscience Project, Ministry of Education, Culture, Sports, Science and Technology, Japan and by a grant of supercomputing time at IDRIS Orsay, France (project number 031393).

References

[1] P. Hohenberg and W. Kohn, Phys. Rev. B **136,** 864 (1964).

[2] W. Kohn and L. J. Sham, Phys. Rev. A **140**, 1133 (1965).

[3] See, e.g., J. P. Perdew, K. Burke, and M. Ernzerhof, Phys. Rev. Lett. **77**, 3865 (1996) and references therein.

[4] H. Yasuhara, S. Yoshinaga and M. Higuchi, Phys. Rev. Lett. **83**, 3250 (1999).

[5] R. Maezono, M. D. Towler, Y. Lee, and R. J. Needs, Phys. Rev. B **68**, 165103 (2003).

[6] See, e.g., W. E. Pickett, Rev. Mod. Phys. **62**, 433 (1989).

[7] A. Fujimori, F. Minami, and S. Sugano, Phys. Rev. B **29**, 5225 (1984).

[8] G. A. Sawatzky and J. W. Allen, Phys. Rev. Lett. **53,** 2339 (1984).

[9] L. Hedin, Phys. Rev. A **139**, 796 (1965); L. Hedin and S. Lundqvist, *Solid State Physics* vol. 23, eds. H. Ehrenreich, F. Seitz, and D. Turnbull (Academic, New York, 1969).

[10] F. Aryasetiawan and O. Gunnarsson, Rep. Prog. Phys. **61**, 237 (1998).

[11] W. G. Aulbur, L. Jönsson, and J. W. Wilkins, Solid State Phys. **54**, 1 (2000).

[12] W. Ku, A. G. Eguiluz, and E. W. Plummer, Phys. Rev. Lett. **85**, 2410 (2000); H. Yasuhara, S. Yoshinaga, and M. Higuchi, Phys. Rev. Lett. **85**, 2411 (2000).

[13] F. Aryasetiawan, Phys. Rev. B **46**, 13051 (1992).

[14] F. Aryasetiawan and O. Gunnarsson, Phys. Rev. Lett. **74**, 3221 (1995).

[15] S. V. Faleev, M. van Schilfgaarde, and T. Kotani, unpublished results.

[16] V. I. Anisimov, J. Zaanen, and O. K. Andersen, Phys. Rev. B **44**, 943 (1991).

[17] V. I. Anisimov, I. V. Solovyev, M. A. Korotin, M. T. Czyzyk, and G. A. Sawatzky, Phys. Rev. B **48**, 16929 (1993).

[18] A. I. Lichtenstein, J. Zaanen, and V. I. Anisimov, Phys. Rev. B **52**, R5467 (1995).

[19] For reviews, see V. I. Anisimov, F. Aryasetiawan, and A. I. Lichtenstein, J. Phys.: Condens. Matter **9**, 767 (1997).

[20] For reviews, see A. Georges, G. Kotliar, W. Krauth, and M. J. Rosenberg, Rev. Mod. Phys. **68**, 13 (1996); T. Pruschke, M. Jarrel, and J. K. Freericks, Adv. Phys. **44**, 187 (1995).

[21] For reviews, see *Strong Coulomb Correlations in Electronic Structure Calculations*, edited by V. I. Anisimov, Advances in Condensed Material Science (Gordon and Breach, New York, 2001).

[22] A. I. Lichtenstein and M. I. Katsnelson, Phys. Rev. B **57**, 6884 (1998).

[23] S. Biermann, F. Aryasetiawan, and A. Georges, Phys. Rev. Lett. **90**, 086402 (2003).

[24] P. Sun and G. Kotliar, Phys. Rev. B **66**, 085120 (2002).

[25] For related ideas, see G. Kotliar and S. Savrasov, in *New Theoretical Approaches to Strongly Correlated Systems,* edited by A. M. Tsvelik (Kluwer Dordrecht, 2001) (and the updated version: cond-mat/0208241).

[26] L. Hedin, Int. J. Quantum Chem. **54**, 445 (1995).

[27] T. Kotani and M. van Schilfgaarde, Solid State Commun. **121**, 461 (2002).

[28] F. Aryasetiawan, M. Imada, A. Georges, G. Kotliar, S. Biermann, and A. I. Lichtenstein, cond-mat/0401620

[29] Q. Si and J. L. Smith, Phys. Rev. Lett. **77**, 3391 (1996).

[30] A. M. Sengupta and A. Georges, Phys. Rev. B **52**, 10295 (1995).

[31] G. Kotliar and H. Kajueter, unpublished results.

[32] H. Kajueter, Ph.D. Thesis, Rutgers University, 1996.

[33] C.-O. Almbladh, U. von Barth and R. van Leeuwen, Int. J. Mod. Phys. B **13**, 535 (1999).

[34] R. Chitra and G. Kotliar, Phys. Rev. B **63**, 115110 (2001).

[35] S. Savrasov and G. Kotliar, cond-mat/0106308.

[36] S. Savrasov, G. Kotliar and E. Abrahams, Nature (London) **410**, 793 (2000).

[37] B. Holm and U. von Barth, Phys. Rev. B **57**, 2108 (1998).

[38] A. I. Lichtenstein, M. I. Katsnelson and G. Kotliar, Phys. Rev. Lett. **87**, 067205 (2001).

[39] O. K. Andersen, Phys. Rev. B **12**, 3060 (1975); O. K. Andersen, T. Saha-Dasgupta, S. Erzhov, Bull. Mater. Sci. **26**, 19 (2003).

[40] M. Springer and F. Aryasetiawan, Phys. Rev. B **57**, 4364 (1998).

[41] J. Bünemann, F. Gebhard, T. Ohm, R. Umstatter, S. Weiser, W. Weber, R. Claessen, D. Ehm, A. Harasawa, A. Kakizaki, A. Kimura, G. Nicolay, S. Shin, V. N. Strocov, Europhys. Lett. **61**, 667 (2003).

[42] H. Mårtensson and P. O. Nilsson, Phys. Rev. B **30**, 3047 (1984).

[43] Y. Motome and G. Kotliar, Phys. Rev. B **62**, 12800 (2000).

[44] J. K. Freericks, M. Jarrell and D. J. Scalapino, Phys. Rev. B **48**, 6302 (1993).

[45] G. Onida, L. Reining and A. Rubio, Rev. Mod. Phys. **74**, 601 (2002).

2 A Many-body Approach to the Electronic and Optical Properties of Copper and Silver

Andrea Marini

2.1 Introduction

Quasiparticles, plasmons and excitons are the fundamental quantities used to interpret the electronic and optical properties of solids. More important than their isolated description is the comprehension of their mutual interaction. However, due to the high complexity and large computational requirements of many-body calculations the experimental band-structures, optical absorption and electron-energy-loss spectra are often compared with the results of (simpler) calculations performed within density functional theory (DFT). The consequences of this approach must however be considered with great care. DFT is based on the idea that the ground state spatial density of a system of interacting particles can be exactly described by a non-interacting gas of fictitious independent particles, moving under the action of an effective potential. Thus DFT, as a ground state theory, makes the comparison with experiments inadequate, in particular for the description of excited states properties. An alternative approach is time-dependent DFT (TDDFT) where all neutral excitations are, in principle, exactly described [1]. However only very recently an efficient approximation for the exchange-correlation kernel of TDDFT has been proposed [2], mainly focused on semiconductors and insulators. The shortcomings of DFT are particularly evident in noble metals where the electronic and optical properties are only qualitatively described: (a) compared with experiment, the usual local density (LDA) approximation to the exchange-correlation potential yields an overestimated d-band width and a too small binding energy for the d-bands top; (b) the experimental electron–energy loss spectrum of silver is dominated by a sharp plasmon peak at 3.83 eV, underestimated in position and almost completely damped when calculated within the DFT-LDA random-phase approximation (RPA); (c) the experimental optical spectra intensity of copper and silver are overall overestimated by ∼30% when calculated within RPA.

In the following sections I will show how these deficiencies of DFT can be successfully corrected using many-body perturbation theory (MBPT). In Section 2.2 I will present the quasiparticle band structure of copper. The effect of the quasiparticle corrections on the optical absorption and EELS of silver will be described in Section 2.3 while in Section 2.4 I will review the most recent results concerning excitonic effects on the optical spectra of metals.

Correlation Spectroscopy of Surfaces, Thin Films, and Nanostructures. Edited by Jamal Berakdar, Jürgen Kirschner
Copyright © 2004 Wiley-VCH Verlag GmbH & Co. KGaA, Weinheim
ISBN: 3-527-40477-5

2.2   Quasiparticle Electronic Structure of Copper

The electronic properties of solids are routinely calculated within DFT in the LDA, by expanding the Kohn–Sham orbitals (KS) in plane waves, see e.g. Ref. [3]. This is made possible by the use of modern norm–conserving pseudopotentials, which allow one to obtain highly accurate valence and conduction band energies without explicitly including the core electrons in the calculation. "Freezing" the core electrons is crucial when a plane-wave basis is used: the number of basis functions needed to describe the 1s of the Si atom is, in fact, 1000 times larger than in the case of the valence shell. The study of noble metals like copper using first-principles methods based on plane-waves and ab-initio pseudopotentials (PPs) presents some peculiar complication with respect to the case of simple metals or semiconductors. In fact, in addition to metalicity, which implies the use of an accurate sampling of the Brillouin zone in order to describe properly the Fermi surface, one must also take into account the contribution of d-electrons to the bonding and to the valence bandstructure. This means that, within the PP scheme, d states cannot be frozen into the core part, but must be explicitly included in the valence, yielding a large total number of valence electrons (11 for bulk copper) [4]. Using soft Martins–Troullier [5] PPs it is possible to work at full convergence with a reasonable kinetic energy cutoff (60 Ry if the 3s and 3p atomic states are frozen into the core, 160 Ry when they are explicitly included) [4]. In Figure 2.1 the DFT band-structure (dashed line) of bulk copper is compared with the experimental data (circles). In contrast to the case of semiconductors, the disagreement between theory and experiment is far from being limited to a rigid shift of the Kohn–Sham occupied eigenvalues with respect to the empty ones. In particular, the comparison clearly shows substantial differences with respect to the experiment for both the d-band width (3.70 instead of 3.17 eV) and position (more than 0.5 eV up-shifted in the DFT), in agreement with previous results [6,7]. The reason for these important deviations of the DFT band-structure is the approximate inclusion of exchange and correlation effects in the LDA single-particle Kohn–Sham (KS) potential. Moreover, in DFT, KS eigenvalues cannot be identified with electron addition or removal energies, since there is no equivalent of Koopman's theorem. MBPT represents an exact method to correct the DFT single particle levels, as the band energies can be obtained in a rigorous way, i.e. as the poles of the one-particle Green's function $G(\boldsymbol{r}, \boldsymbol{r}'; \omega)$ [8]. The latter are determined by an equation of the form:

$$\left[-\frac{\hbar^2}{2m} \triangle_{\boldsymbol{r}} + V_{\text{external}}(\boldsymbol{r}) + V_{\text{Hartree}}(\boldsymbol{r}) \right] \psi_{n\boldsymbol{k}}(\boldsymbol{r}, \omega)$$

$$+ \int \mathrm{d}\boldsymbol{r}' \, \Sigma(\boldsymbol{r}, \boldsymbol{r}'; \omega) \, \psi_{n\boldsymbol{k}}(\boldsymbol{r}', \omega) = E_{n\boldsymbol{k}}(\omega) \, \psi_{n\boldsymbol{k}}(\boldsymbol{r}, \omega), \quad (2.1)$$

containing the non-local, non-hermitian and frequency dependent self-energy operator Σ. The poles of G are the QP energies $\epsilon_{n\boldsymbol{k}}^{\text{QP}}$, the solutions of $\epsilon_{n\boldsymbol{k}}^{\text{QP}} = E_{n\boldsymbol{k}}\left(\epsilon_{n\boldsymbol{k}}^{\text{QP}}\right)$. The corresponding quasiparticle wavefunctions are $\phi_{n\boldsymbol{k}}(\boldsymbol{r}) = \psi_{n\boldsymbol{k}}\left(\boldsymbol{r}, \epsilon_{n\boldsymbol{k}}^{\text{QP}}\right)$. The off-diagonal matrix elements of the self-energy operator $\langle n\boldsymbol{k}|\Sigma(\boldsymbol{r}_1, \boldsymbol{r}_2, \omega)|n'\boldsymbol{k}'\rangle$ are usually much smaller than the diagonal elements. Thus the quasiparticle and the KS wavefunctions can be assumed identical and

Eq. (2.1) can be rewritten as a scalar equation for $\epsilon_{nk}^{\mathrm{QP}}$,

$$\epsilon_{nk}^{\mathrm{QP}} = \epsilon_{nk}^{\mathrm{DFT}} + M_{nk}\left(\epsilon_{nk}^{\mathrm{QP}}\right) + \Sigma_x^{nk} - V_{xc}^{nk}. \tag{2.2}$$

Following Ref. [9], we have separated the static, bare-exchange part $\Sigma_x\left(r, r'\right)$ from $M\left(r, r'; \omega\right)$, the energy-dependent correlation contribution, or mass-operator. Σ_x^{nk} is given by:

$$\Sigma_x^{nk} = -\sum_{n_1} \int_{\mathrm{BZ}} \frac{dq}{(2\pi)^3} f_{n_1(k-q)}$$
$$\iint dr\, dr'\, \phi_{nk}^*\left(r\right) \phi_{n_1(k-q)}\left(r\right) v\left(r, r'\right) \phi_{n_1(k-q)}^*\left(r'\right) \phi_{nk}\left(r'\right), \tag{2.3}$$

where $v\left(r, r'\right)$ is the bare Coulomb interaction and $0 \le f_n\left(k\right) \le 1$ represents the occupation number. M is usually evaluated according to the so-called GW approximation, derived by Hedin in 1965 [8, 10], which is based on an expansion in terms of the dynamically screened Coulomb interaction $W\left(r, r'; \omega\right)$:

$$M_{nk}\left(\omega\right) = \langle nk| M\left(r_1, r_2, \omega\right) |nk\rangle$$
$$= -\sum_{n_1} \int_{\mathrm{BZ}} \frac{dq}{(2\pi)^3} \int_{-\infty}^{\infty} d\omega' \left[\frac{\Gamma_{nn_1}^v\left(k, q, \omega'\right)}{\omega - \omega' - \epsilon_{n_1(k-q)}^{\mathrm{DFT}} + i\delta} \right.$$
$$\left. + \frac{\Gamma_{nn_1}^c\left(k, q, \omega'\right)}{\omega - \omega' - \epsilon_{n_1(k-q)}^{\mathrm{DFT}} - i\delta} \right]. \tag{2.4}$$

Γ^c and Γ^v are the conduction, valence contributions to the self-energy spectral function:

$$\Gamma_{nn_1}^c\left(k, q, \omega\right) = 2\left(1 - f_{n_1(k-q)}\right) \iint dr\, dr'\, \phi_{nk}^*\left(r\right) \phi_{n_1(k-q)}\left(r\right)$$
$$W^\delta\left(r, r'; \omega\right) \theta\left(-\omega\right) \phi_{n_1(k-q)}^*\left(r'\right) \phi_{nk}\left(r'\right), \tag{2.5}$$

$$\Gamma_{nn_1}^v\left(k, q, \omega\right) = 2 f_{n_1(k-q)} \iint dr\, dr'\, \phi_{nk}^*\left(r\right) \phi_{n_1(k-q)}\left(r\right)$$
$$W^\delta\left(r, r'; \omega\right) \theta\left(\omega\right) \phi_{n_1(k-q)}^*\left(r'\right) \phi_{nk}\left(r'\right). \tag{2.6}$$

Here W^δ is the delta–like part of the Lehman representation of the screened Coulomb interaction function. W is expressed in terms of the microscopical inverse dielectric function, $W\left(r, r'; \omega\right) = v\left(r, r'\right) + \int dr'' v\left(r, r''\right) \epsilon^{-1}\left(r'', r'; \omega\right)$.

Most GW calculations on semiconductor systems use a plasmon-pole approximation (PPA) for $W\left(\omega\right)$ [11], based on the observation that the Fourier components of the inverse dielectric function are generally peaked functions of ω, and can be approximated by a single pole. Since the evaluation of M involves an integration over the energy, the fine details of the ω-dependence are not critical, and the PPA turns out to work reasonably well for most applications. However, in the case of copper, the use of a PPA becomes more critical. The presence of flat d-bands 2 eV below the Fermi level implies strong transitions in the inverse dielectric function spread over a large energy range. These transitions are not at all well described as a single-pole function, leading to instabilities when determining the plasmon-pole parameters. Instead, the screened electron–hole interaction must be explicitly computed over a grid of about 200 frequencies from zero to ~130 eV, and the energy integral performed numerically.

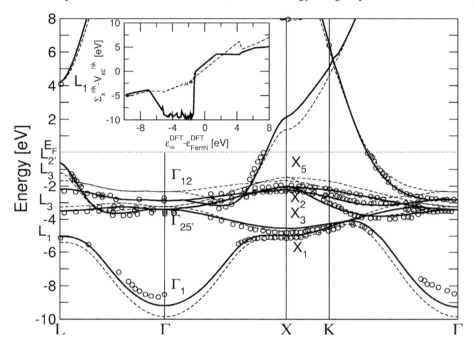

Figure 2.1: Full line: quasiparticle GW results for the bulk copper bandstructure [12], compared with the DFT-LDA results (dashed line), and with the experimental data reported in Ref. [7] (circles). Inset: values of $\Sigma_x^{nk} - V_{xc}^{nk}$, plotted as a function of the non-interacting energies ϵ_{nk}^{DFT}. The dashed line represents the results obtained without the contributions from the 3s and 3p core states.

The set of Eqs. (2.1)–(2.6) constitutes a well-defined and successful scheme to calculate quasiparticle band structure in many different materials [8]. Normally (e.g., in GW calculations for semiconductors), the calculation of G and W to correct the DFT valence bandstructure can be performed by including only valence states, and fully neglecting the core states which have been frozen in the pseudopotential approach. Among transition metals, full quasiparticle calculations have been carried out only for Ni [13]. In the case of Ni, GW yields a good description of photoemission data, except for the 6 eV satellite, which is due to strong short-range correlations within the partially filled d-shell. For copper, we would expect band-theory to work better than for transition metals, since d-shells are completely filled. Instead, as showed in Ref. [12], when Σ is computed neglecting the 3s and 3p atomic core states (which in the solid create two flat bands, at about 112 and 70 eV, respectively, below the Fermi level), the resulting QP corrections on the d-bands are clearly unphysical: GW corrections move the highest occupied d-bands above the DFT-LDA Fermi level. On the other hand, the situation for s/p states (e.g. for the state $L_{2'}$) is much more reasonable, with correlation and exchange parts of the self-energy which largely cancel each other (as in the case of semiconductors), and negative QP corrections of the order of eV. The solution of this puzzling situation is provided by the role of the above-mentioned 3s and 3p states, which, despite being well separated *in energy* from the 3d ones, have a large *spatial* overlap with the latter. As a consequence, non-negligible contributions to the self-energy are expected from the *exchange* contributions between 3d and 3s/3p states as clearly shown in the inset of Figure 2.1, where the difference $\Sigma_x^{n\mathbf{k}} - V_{xc-\mathrm{LDA}}^{n\mathbf{k}}$ is strongly affected by the presence of core levels in the n_1 summation of Eq. (2.4). The role of core levels in the calculation of the bare exchange contributions (whose importance has already been addressed for transition metals by Aryasetiawan and Gunnarsson [8], but estimated to be of the order of 1 eV) is hence crucial in the case of copper. Moreover their effect is unexpected on the basis of DFT-LDA calculations, where the band structure does not change appreciably even if the 3s/3p orbitals are fully included in the valence [4].

In Figure 2.1 the full GW theoretical band structure [12] is compared with experimental data: the agreement is remarkably good and the fact that the GW corrections cannot be reproduced by any rigid shift of the LDA bands clearly appears. Hence, the GW method, originally devised to describe the *long-range* charge oscillations [10], is hence shown to yield a good description also of copper, a system characterized by localized orbitals and short-range correlation effects.

2.3   The Plasmon Resonance of Silver

In the previous section we have seen how to correct the DFT single particle levels using MBPT obtaining an excellent agreement with the experimental results. However in a fully interacting electronic system quasiparticles coexist with collective excitations, i.e. plasmons. Plasmons occur at energies for which the real part of the dielectric function vanishes with a corresponding small imaginary part; they can be observed experimentally as sharp peaks in electron energy loss spectra (EELS). A well established technique to calculate EELS uses the independent particles or random phase approximation (RPA) for the polarization function, obtained in terms of ab-initio single-particle energy bands. In the case of silicon, Olevano and Reining [14] showed that using the quasiparticle energies without going beyond the RPA, including excitonic effects (as we will discuss in the next section), the shape of the plasmon peak worsens with respect to the experiment. Inclusion of self-energy corrections and excitonic effects yields a spectrum very similar to the DFT-LDA one and to the experiment. Ku and Eguiluz [15] obtained a correct positive dispersion of the plasmon width in K using the single particle approximation, with no many-body corrections beyond DFT-LDA. In these cases many-body effects (quasiparticle corrections and/or excitonic effects) are not required to describe correctly the experimental data. These results agree with the general feeling that excitonic effects partially cancel self-energy corrections (we will discuss in detail this cancellation in the next section). A similar result has been found for copper [4, 16], where the RPA response function calculated without many-body corrections yields good agreement with the experimental EEL and optical spectra.

In this framework the case of silver is rather surprising: the experimental EELS (circles in Figure 2.2) is dominated by a sharp plasmon peak at 3.83 eV [17], whose position and width are badly reproduced in DFT-LDA RPA [18] (dashed line in Figure 2.2). In particular, a width of about 0.5 eV is obtained within this approach, to be contrasted with a much narrower experimental width (~100 meV). A similar discrepancy occurs in the reflectance spectrum (see Figure 2.3), where a very narrow dip at 3.92 eV is hardly reproduced by DFT-LDA calculations.

Following the same GW scheme used for copper in the previous section the quasiparticle band structure of silver [19] at high symmetry points is compared with DFT-LDA results in Table 2.1. While the deeper energy levels remain mostly unchanged, a downward shift of about 1.3 eV of the top d bands leads to a decrease of the bandwidth, and hence to an excellent agreement with experiment.

Table 2.1: Theoretical band widths and band energies for silver [19], at high-symmetry points. GW energies are relative to the QP Fermi Level. The striking agreement with the experimental results shows that the silver band-structure is very well described at the GW level. The values in the last column are taken from Ref. [20] where spin–orbit splittings have been removed by making degeneracy–weighted averages.

		DFT-LDA	GW	Experiment
Positions	Γ_{12}	−3.57	−4.81	−4.95
of	X_5	−2.49	−3.72	−3.97
d-bands	$L_3(2)$	−2.71	−3.94	−4.15
	$\Gamma_{12} - \Gamma_{25'}$	1.09	0.94	1.11
Widths	$X_5 - X_3$	3.74	3.39	3.35
of	$X_5 - X_1$	3.89	3.51	3.40
d-bands	$L_3(2) - L_3(1)$	1.98	1.85	1.99
	$L_3 - L_1$	3.64	3.17	2.94
	$X_5 - X_2$	0.27	0.29	0.38

To calculate the EEL spectra, the most simple expression for the dielectric function is obtained within RPA, where the electron and holes excited by the external perturbation are assumed to move independently. The EELS is given by the imaginary part of the inverse dielectric function $\epsilon^{-1}(\omega)$:

$$\epsilon^{-1}(\omega) = \left[\epsilon_{\mathrm{ib}}(\omega) - \frac{\omega_{\mathrm{D}}^2}{\omega(\omega + i\eta)}\right]^{-1}, \tag{2.7}$$

where $\epsilon_{\mathrm{ib}}(\omega)$ is the interband contribution and $\omega_{\mathrm{D}} = 9.48$ eV is the Drude plasma frequency, both calculated ab-initio following the procedure described in Ref. [4]. The interband RPA dielectric function is given by

$$\epsilon_{\mathrm{ib}}\left(\omega\right)=1-4\pi\lim_{q\to0}\int_{\mathrm{BZ}}\frac{d^{3}k}{\left(2\pi\right)^{3}}\sum_{n\neq n'}\frac{\left|\left\langle n'k-q|\mathrm{e}^{-iq\cdot r}|nk\right\rangle\right|^{2}}{\left|q\right|^{2}}$$
$$\frac{f_{n',k-q}-f_{n,k}}{\omega+E_{n,k}-E_{n',k-q}+i\eta}, \quad (2.8)$$

where $\langle n'k-q|\mathrm{e}^{-iq\cdot r}|nk\rangle = \int \mathrm{d}r\mathrm{e}^{-iq\cdot r}\phi_{n'k-q}^{*}\left(r\right)\phi_{nk}\left(r\right)$. The $q\to0$ limit of Eq. (2.8) has been done including the effects of the pseudopotential non-locality, as described in Ref. [4]. Using as single-particle energies E_{nk} the KS levels we obtain the DFT-LDA RPA dielectric function, showed in the inset of Figure 2.2 (dashed line). Because of the under-estimation of the d-bands top position (see Table 2.1) the interband onset is too low compared to the experiment (circles) [21]. Therefore the corresponding plasmon peak is strongly damped and its energy underestimated (dashed line in the main frame of Figure 2.2). When the quasiparticle single-particle energies are used in Eq. (2.8) instead of the DFT-LDA ones, the plasmon peak, underestimated in intensity and position in DFT-LDA, is shifted toward higher energies and strongly enhanced by GW corrections, in striking agreement with ex-periment. This resonance can be interpreted as a collective (Drude-like) motion of electrons in the partially filled band. However, its energy ω_{p} does not coincide with the bare Drude frequency ω_{D}, the difference arising from the screening of the electron–electron interaction by virtual interband transitions. The plasmon resonance, although blue shifted with respect to DFT-LDA, remains *below* the main interband threshold, but overlaps the weak low-energy tail of interband transitions acquiring a small, yet finite, width.

The polarization of the "medium" where the plasmon oscillates (the d electrons) is hence important to determine its energy and width. This polarization is absent in the homogeneous electron gas because there are no localized d orbitals and no interband transitions; it is weak in semiconductors (like Si), because interband transitions occur at energies far from that of the plasma resonance. The same polarization effect is present, but destructive in copper due to the lower onset of interband transitions. EELS peaks occur *above* this onset and are therefore strongly broadened. Consequently, the delicate interplay of plasmon-frequency renormalization with the shift of the interband-transition onset, both due to QP corrections, may yield (in silver) or may not yield (in copper) a sharp plasmon resonance.

Another important quantity is the reflectance, $R\left(\omega\right) = \left(\left|N\left(\omega\right)-1\right| / \left|N\left(\omega\right)+1\right|\right)^{2}$, where N is the complex refraction index defined by $\left[N\left(\omega\right)\right]^{2} = \epsilon\left(\omega\right)$. In Figure 2.3 we com-pare the GW $R\left(\omega\right)$ with the DFT-LDA one, and with experimental results [22]. The latter shows a very narrow dip at $3.92\,\mathrm{eV}$, close to the plasmon frequency, arising from the zero-reflectance point ω_{0}, defined as $\epsilon\left(\omega_{0}\right) = 1$. Again, the width and depth of this reflectance dip are related to the imaginary part of $\epsilon\left(\omega\right)$. GW corrections make ω_{0} occur below the main on-set of interband transitions, and hence produce a very narrow and deep reflectance minimum. Here the agreement between GW results and experiments for the intensity and width of the dip at $3.92\,\mathrm{eV}$ is even more striking than in the EELS.

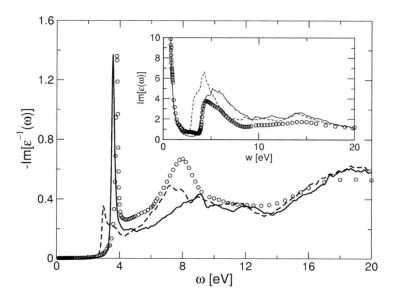

Figure 2.2: Electron energy loss spectrum (EELS) of silver from Ref. [19]. Solid line: GW. Dashed line: DFT-LDA. Circles: experiment [22]. The non trivial quasiparticle GW corrections improve considerably the DFT-LDA plasmon peak, yielding a striking agreement with the experiment. Inset: optical absorption of silver within RPA. Solid line: GW energy levels are used in Eq. (2.8). Dashed line: DFT-LDA energy levels are used. Circles: experiment [22].

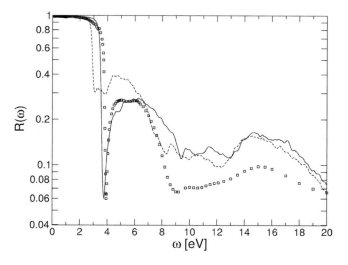

Figure 2.3: Reflectivity spectrum of silver from Ref. [19]. Solid line: GW. Dashed line: DFT-LDA. Boxes: experiment [22]. The experimental sharp dip at 3.92 eV is correctly reproduced by GW, with a substantial improvement on the DFT-LDA spectrum.

2.4 Dynamical Excitonic Effects in Metals

In the inset of Figure 2.2 we observe a strong overestimation of the absorption spectrum intensity with respect to the experiment (similar to that found in copper [4]). However this residual discrepancy cannot be traced back to the standard phenomenology observed in semiconductors and insulators where the observed light-absorption spectra largely deviate from independent-particle, RPA calculations [1]. These deviations are corrected by including the electron–hole interaction [23]. The strength of these modifications increases as the inverse of the dielectric constant of the system. In insulators, where the electron–hole interaction is only weakly screened, sharp peaks with energy below the optical gap (bound excitons) can be observed in the experimental spectra [24]. If this description is extrapolated to the metallic case the natural conclusion is that the electron–hole interaction has a negligible effect on the optical spectra of metals as the static electron–hole interaction is completely screened by the long–range part of the dielectric function. This simple argument has been considered definitive to assert that there are no excitonic effects in metals. However the standard approach to account for the electron–hole interaction in the optical spectra involves approximations whose validity is assumed a priori, and not sustained by theoretical motivations. These approximations are the key to understanding the apparently inexplicable overestimation observed in the RPA optical spectra of silver and copper.

To introduce the theoretical framework commonly used to calculated optical properties beyond the RPA we have to rewrite the dielectric function as $\epsilon(\omega) \equiv 1 - 8\pi \boldsymbol{\Lambda}^{\dagger} \boldsymbol{P}(\omega) \boldsymbol{\Lambda}$, where $\boldsymbol{P}(\omega)$ is the matrix representation of the polarization function in the non-interacting electron–hole basis and $\boldsymbol{\Lambda}$ is a vector embodying the corresponding optical oscillators $\langle n'\boldsymbol{k} - \boldsymbol{q}|e^{-i\boldsymbol{q}\cdot\boldsymbol{r}}|n\boldsymbol{k}\rangle$ already introduced in Eq. (2.8). The polarization function is obtained by solving the Bethe–Salpeter equation (BSE), an integral equation for the four point electron–hole Green's function $\boldsymbol{L}(t_1, t_2; t_3, t_4)$ [1, 23, 25]. As we are interested in the polarization function $\boldsymbol{P}(t) \equiv -i\boldsymbol{L}(t, 0; t, 0)$, the BSE can be rewritten as:

$$
\boldsymbol{P}(t) = \boldsymbol{P}^{(0)}(t) - \int dt_1 \, \boldsymbol{P}^{(0)}(t - t_1) \, \boldsymbol{V} \boldsymbol{P}(t_1)
$$
$$
+ \iint dt_1 dt_2 \, \boldsymbol{L}^{(0)}(t, t_2; t, t_1) \, \widetilde{\boldsymbol{W}}(t_1 - t_2) \, \boldsymbol{L}(t_1, 0; t_2, 0). \quad (2.9)
$$

Equation (2.8) is obtained when no electron–hole effects are included in $\boldsymbol{P}$, i.e. $\boldsymbol{P} \approx \boldsymbol{P}^{(0)} = -i\boldsymbol{L}^{(0)}$. Like $\boldsymbol{P}$, also $\boldsymbol{V}$, $\boldsymbol{L}$ and $\widetilde{\boldsymbol{W}}$ are matrix representations in the non-interacting electron–hole basis. Using generalized indexes $\boldsymbol{K} := (c\,v\,\boldsymbol{k})$; $c, v, \boldsymbol{k}$ being conduction, valence band and k-point indexes, we can express those quantities as:

$$
\widetilde{W}_{\boldsymbol{K}\boldsymbol{K}'}(t_1 - t_2) = i\langle c\boldsymbol{k}, v\boldsymbol{k}|W(\boldsymbol{r}, \boldsymbol{r}'; t_1 - t_2) - v(\boldsymbol{r}, \boldsymbol{r}')|c'\boldsymbol{k}', v'\boldsymbol{k}'\rangle, \quad (2.10)
$$

$$
V_{\boldsymbol{K}\boldsymbol{K}'}(t_1 - t_2) = i \iint d\boldsymbol{r}d\boldsymbol{r}' \phi_{c\boldsymbol{k}}^*(\boldsymbol{r}) \phi_{v\boldsymbol{k}}(\boldsymbol{r}) v(\boldsymbol{r}, \boldsymbol{r}') \phi_{v'\boldsymbol{k}'}^*(\boldsymbol{r}') \phi_{c'\boldsymbol{k}'}(\boldsymbol{r}'). \quad (2.10')
$$

$L^{(0)}$ is the non-interacting electron–hole Green's function

$$L^{(0)}_{\boldsymbol{K},\boldsymbol{K}'}(t_1,t_2;t_3,t_4) = \delta_{v,v'}\delta_{c,c'}\delta_{\boldsymbol{k},\boldsymbol{k}'} Z_{c\boldsymbol{k}} Z_{v\boldsymbol{k}}$$
$$\theta(t_1-t_4)\,\mathrm{e}^{-iE_{c\boldsymbol{k}}(t_1-t_4)}\theta(t_3-t_2)\,\mathrm{e}^{iE_{v\boldsymbol{k}}(t_3-t_2)}. \quad (2.11)$$

$Z_{n\boldsymbol{k}}$ (smaller than 1) are the QP energies and renormalization factors, respectively. The latter represent the weights of the QP peak in the many-body single-particle spectral function. The more $Z_{n\boldsymbol{k}}$ differs from 1, the more the high energy structures in the spectral function (like plasmonic replicas) become important. Those high-energy peaks are not visible in the optical energy range but, nevertheless, subtract intensity from the QP peaks. Very little is known about the role played by the Z factors in optical spectra calculations [26]. In Section 2.3 we had to assume $Z = 1$, because if the Z factors are included in an RPA calculation [27], or even in the BSE (see below), the intensity of the resulting spectra is strongly underestimated, both in metals and in semiconductors. Thus the Z factors are commonly set to 1 *by hand* in the solution of the BSE or in the calculation of the independent-QP spectra. This is the first approximation needed to reproduce the experimental results, and it is important to note that it lacks sound theoretical justification.

But there is another important approximation needed to solve Eq. (2.9), related to the time dependent term $\widetilde{\boldsymbol{W}}(t_1 - t_2)$. Indeed, the BSE with a time-dependent interaction is considered hardly solvable (if not "practically unsolvable" [26]) and *for computational convenience* the electron–hole interaction is assumed to be instantaneous; this is equivalent to approximating W with its static value, $W(\boldsymbol{r}_1, \boldsymbol{r}_2, \omega = 0)$. This approximation is verified a posteriori through comparison with experiment and physically corresponds to the assumption that the electron–hole scattering time is much longer than the characteristic screening time of the system (roughly speaking, the inverse of the plasma frequency). Indeed, the static approximation is expected to work well for transition energies much smaller than the plasma frequency [23]. However the most striking examples of systems that do not fulfill this condition are silver and copper. From Figure 2.2 it is evident that the plasmon of silver, which dominates the EELS, is located just above the interband gap ($\sim 3.9\,\mathrm{eV}$). Similarly, the EELS of copper shows strong, broad peaks in the optical range [4]. When $\widetilde{\boldsymbol{W}}(t) \approx \widetilde{\boldsymbol{W}}(\omega = 0)\,\delta(t)$ [1,23]. Equation ((2.9)) can be formally solved by means of a Fourier transform:

$$\boldsymbol{P}(\omega) = \boldsymbol{P}^{(0)}(\omega) - \boldsymbol{P}^{(0)}(\omega)\left(\boldsymbol{V} + \widetilde{\boldsymbol{W}}\right)\boldsymbol{P}(\omega). \quad (2.12)$$

This is the static BSE (SBSE) commonly applied neglecting the renormalization factors in Eq. (2.9), i.e. taking $Z_{n\boldsymbol{k}} = 1$. It yields optical spectra in good agreement with experiments in semiconductors and insulators [23]. When applied to copper and silver however, the SBSE result (dotted lines in Figure 2.5) is indistinguishable from the independent-QP calculation, without improving the agreement with experiment.

In Ref. [28] a solution of Eq. (2.9) is proposed without the two major approximations employed in the SBSE, i.e. keeping the $Z_{n\boldsymbol{k}}$ factors smaller than 1 and W frequency dependent. To this end $L^{(0)}\widetilde{\boldsymbol{W}}L$ is expanded in powers of $\widetilde{\boldsymbol{W}}$. The first order term of this expansion,

$P^{(1)}(t)$, is given by:

$$P^{(1)}_{\mathbf{K}_1 \mathbf{K}_2}(t) = \iint \mathrm{d}t_1 \, \mathrm{d}t_2 \theta (t_1 - t_2)$$

$$\left[L^{(0)}_{\mathbf{K}_1}(t, t_2; t, t_1) \, \widetilde{W}_{\mathbf{K}_1 \mathbf{K}_2}(t_1 - t_2) \, L^{(0)}_{\mathbf{K}_2}(t_1, 0; t_2, 0) \right.$$

$$\left. + L^{(0)}_{\mathbf{K}_1}(t, t_1; t, t_2) \, \widetilde{W}_{\mathbf{K}_1 \mathbf{K}_2}(t_2 - t_1) \, L^{(0)}_{\mathbf{K}_2}(t_2, 0; t_1, 0) \right]. \quad (2.13)$$

From Eq. (2.10) it is straightforward to see that

$$L^{(0)}_{\mathbf{K}_1}(t, t_2; t, t_1) = i \left[P^{(0)}_{\mathbf{K}_1}(t - t_1) \, \mathrm{e}^{iE_{v_1 k_1}(t_1 - t_2)} \theta (t_1 - t_2) \right.$$

$$\left. + P^{(0)}_{\mathbf{K}_1}(t - t_2) \, \mathrm{e}^{-iE_{c_1 k_1}(t_2 - t_1)} \theta (t_2 - t_1) \right], \quad (2.14)$$

$$L^{(0)}_{\mathbf{K}_2}(t_1, 0; t_2, 0) = i \left[P^{(0)}_{\mathbf{K}_2}(t_2) \, \mathrm{e}^{-iE_{c_2 k_2}(t_1 - t_2)} \theta (t_1 - t_2) \right.$$

$$\left. + P^{(0)}_{\mathbf{K}_2}(t_1) \, \mathrm{e}^{iE_{v_2 k_2}(t_2 - t_1)} \theta (t_2 - t_1) \right], \quad (2.14')$$

that inserted in Eq. (2.13), casts $P^{(1)}(t)$ as a time convolution of three terms (as shown diagrammatically in Figure 2.4).

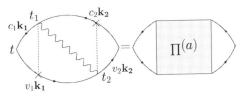

Figure 2.4: Diagrammatic representation of the first order contribution to the polarization function $P(t)$ according to the BSE. Crosses indicate the time points where the incoming and outgoing non-interacting Green's functions are "cut" according to Eqs. (2.14). The right-hand side represents the first order polarization diagram of the dynamical Bethe–Salpeter as discussed in the text. It can be used to sum all orders of BSE with non overlapping interaction lines.

As a consequence, in the frequency domain $P^{(1)}(\omega)$ has the form:

$$\boldsymbol{P}^{(1)}(\omega) = -\boldsymbol{P}^{(0)}(\omega) \left[\boldsymbol{\Pi}^{(a)}(\omega) + \boldsymbol{\Pi}^{(b)}(\omega) \right] \boldsymbol{P}^{(0)}(\omega), \quad (2.15)$$

with

$$\Pi^{(a)}_{\mathbf{K}_1 \mathbf{K}_2}(\omega) = \widetilde{W}^{(+)}_{\mathbf{K}_1 \mathbf{K}_2}(\omega + E_{v_1 k_1} - E_{c_2 k_2}), \quad (2.16)$$

and

$$\Pi^{(b)}_{\mathbf{K}_1 \mathbf{K}_2}(\omega) = \widetilde{W}^{(+)}_{\mathbf{K}_1 \mathbf{K}_2}(\omega + E_{v_2 k_2} - E_{c_1 k_1}); \quad (2.16')$$

$\widetilde{W}^{(+)}(\omega)$ being the Laplace transform of $\widetilde{W}(t)$. The two terms denoted by (a) and (b) correspond to the two possible time orderings of the interaction ends ($t_1 > t_2$ for term (a), shown in Figure 2.4; $t_2 > t_1$ for term (b), not shown). Equation (2.15) can be thought of as the first order expansion of $P(\omega)$ in the frequency–dependent interaction $\Pi(\omega) = \Pi^{(a)}(\omega) + \Pi^{(b)}(\omega)$, which replaces $\widetilde{W}$ of the SBSE. Thus a partial summation of the BSE can be performed writing:

$$P(\omega) = P^{(0)}(\omega) - P^{(0)}(\omega)\left[V + \Pi(\omega)\right]P(\omega). \tag{2.17}$$

This is the Dynamical Bethe–Salpeter equation (DBSE) [28]. The diagrams summed up in Eq. (2.17) are those containing the ladder series of repeated electron–hole interactions with *non overlapping* (in time) interaction lines. The poles of $P(\omega)$, Ω_λ, will be given by the solution of the equation $\left[P^{(0)}(\Omega_\lambda)\right]^{-1} + V + \Pi(\Omega_\lambda) = 0$. In contrast to the kernel of the SBSE, $\Pi(\Omega_\lambda)$ is not hermitian and, consequently, Ω_λ is in general complex. Its imaginary part gives the inverse excitonic lifetime. Thus the interacting electron–hole states are actually dressed excitons, or *quasiexcitons*. This agrees with what has been already found in the core exciton limit [29] and emphasizes the analogy between the DBSE and the Dyson equation. Consequently, as in the single-particle problem, we expect to find similar renormalization effects on the quasiexcitonic Green's function. To develop further this aspect we expand linearly the smooth function $\widetilde{W}^{(+)}(\omega)$ around the non-interacting electron–hole energies, obtaining $\Pi_{K_1 K_2}(\omega) \approx \Pi_{K_1 K_2}^{(\mathrm{st})} + \Theta_{K_1 K_2}(\omega - E_{c_2 k_2} + E_{v_2 k_2})$. $\Pi_{K_1 K_2}^{(\mathrm{st})} = \left.\Pi_{K_1 K_2}(\omega)\right|_{\omega = E_{c_2 k_2} - E_{v_2 k_2}}$ is the static limit of the dynamical Bethe–Salpeter kernel which turns out to be quite similar to the kernel of the SBSE. $\Theta_{K_1 K_2} = \left.\partial\Pi_{K_1 K_2}(\omega)/\partial\omega\right|_{\omega = E_{c_2 k_2} - E_{v_2 k_2}}$ are the excitonic dynamical-renormalization factors. Thus Eq. (2.17) can be strongly simplified in the case of copper and silver where the effect of $\Pi^{(\mathrm{st})} + V$ is very small. The corresponding polarization function $P(\omega)$ is approximatively given by:

$$P_{K_1 K_2}(\omega) \approx \frac{\left[\left(Z^{\mathrm{eh}}\right)^{-1} + \Theta\right]^{-1}_{K_1 K_2}}{\omega - E_{c_2 k_2} + E_{v_2 k_2} + i0^+}, \tag{2.18}$$

with $Z^{\mathrm{eh}}_{K_1 K_2} = Z_{c_1 k_1} Z_{v_1 k_1} \delta_{K_1 K_2}$.

The connection between dynamical excitonic and self-energy effects is now clear. $Z_{nk}^{-1} = 1 - \beta_{nk}$, where the negative factor β_{nk}, the frequency derivative of the self-energy, is the weight lost by the QP because of the coupling with the excitations of $W(\omega)$. The excitonic factors Θ, instead, are due to the modification of such coupling as a consequence of the electron–hole interaction. Those two effects tend to cancel each other but *the cancellation is, in general, not complete*, as exemplified in Figure 2.5 for copper and silver. The SBSE calculation (dotted line), with $Z_{nk} = 1$ and $\Theta = 0$, overestimates the experimental intensity (circles), while the inclusion of the Z_{nk} factors only (dashed line) underestimates it. In the DBSE (full line) the dynamical Θ factors partially compensate for the Z^{eh} factors yielding a spectral intensity in good agreement with experiment.

Similarly the optical spectra of semiconductors can be obtained with the DBSE in excellent agreement with experiment, as shown in the case of silicon in Ref. [28]. In contrast to the metallic case, however, the DBSE kernel of silicon must contain second-order contributions in order to reproduce correctly the experimental optical spectrum. The main effect of the first order kernel $\boldsymbol{\Pi}(\omega)$ is indeed to balance the reduction of optical strengths due to self-energy renormalization factors, as suggested by Bechstedt et al. [26]. However, the renormalized QP weights also imply a reduction of the statically screened electron–hole of almost $\sim 30\%$, which is the reason for the wrong relative intensities of the two peaks in the SBSE result. This shortcoming is fixed by the second-order diagrams.

Thus in both the metallic and semiconducting cases the DBSE correctly describes the measured optical spectra without the a priori approximations commonly used in solving the BSE. This result can be interpreted by thinking of the electron–hole pair as a neutral excitation, thus, less efficient than the electron and the hole alone in exciting virtual plasmons, which is the main process leading to QP renormalization. Only when dynamical effects are coherently included both in the self energy and in the electron–hole interaction does this (physically expected) result emerges from the bundle of many-body equations. *This confirms the SBSE results but not the separate approximations involved therein.*

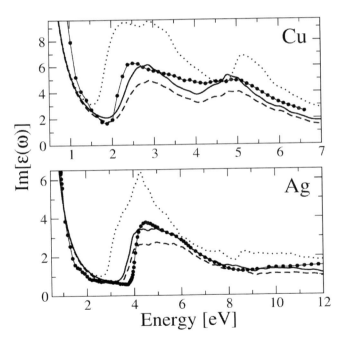

Figure 2.5: Absorption spectrum of bulk copper and silver from Ref. [28]. Dotted line: SBSE without renormalization factors. Dashed line: SBSE including the QP renormalization factors. Full line: result of the dynamical Bethe–Salpeter equation including dynamical QP and excitonic effects. Circles: experimental spectra (Ref. [22]).

2.5   Conclusions

Several conclusions can be drawn from the presented results. Starting from DFT we have reviewed all the steps needed to correct the observed disagreements between the experiment and the calculated band-structure (Section 2.2), electron energy loss spectrum (Section 2.3) and optical spectra (Section 2.4). We have observed in all cases a key role played by the localized d-orbitals. They contribute to the self-energy with strong exchange effects with the 3s/3p core states, balancing the large correlation part of the mass-operator. Thus core levels cannot be neglected in the quasiparticle calculations of noble metals. Then, in contrast to semiconductors and simple metals, the plasma frequency of silver is renormalized by the interband transitions between d and s/p bands when QP single particle energies are used to calculate the RPA polarization function. This clearly shows that the low-energy peaks in the dielectric function due to the d states affect also the excitations occurring at the Fermi surface where the main contribution to the metallic screening arises. Finally, even if the s/p metallic bands completely screen the electron–hole interaction at *zero* energy (static contribution), the localized d-orbitals induce a dynamical term in the Bethe–Salpeter kernel that partially cancel the dynamical self-energy effects (renormalization factors). In conclusion single-particle, two-particle excitations, as well as plasmons in copper and silver, can be successfully described within many-body perturbation theory. The role of the localized, d-orbitals must be correctly included in the calculations and, more importantly, in the theory. These unique features of noble and, in general, d metals make them a stringent test for modern ab-initio Many-Body theories.

References

[1] For a review, see G. Onida, L. Reining, and A. Rubio, Rev. Mod. Phys. **74** (2002) 601.

[2] A. Marini, A. Rubio, and R. Del Sole,   Phys. Rev. Lett. **91** (2003) 256402;   F. Sottile, V. Olevano and L. Reining,   Phys. Rev. Lett. **91** (2003) 056402.

[3] O. Pulci et al., Phys. Rev. B **55** (1997) 6685.

[4] A. Marini, G. Onida and R. Del Sole, Phys. Rev. B **64** (2001) 195125.

[5] N. Troullier and J.L. Martins, Phys. Rev. B **43** (1991) 1993.

[6] See, for example, V.N. Strocov et al., Phys. Rev. Lett. **81** (1998) 4943; Phys. Rev. B **63** (2001) 205108.

[7] R. Courths and S. Hüfner, Phys. Rep. **112** (1984) 53.

[8] F. Aryasetiawan and O. Gunnarsson, Rep. Prog. Phys. **61** (1998) 237-312.

[9] R. W. Godby, M. Schlüter and L. J. Sham, Phys. Rev. B **37** (1988) 10159.

[10] L. Hedin, Phys. Rev. A **139** (1965) A796.

[11] M. S. Hybertsen and S. G. Louie, Phys. Rev. B **34** (1986) 5390; R.W. Godby and R.J. Needs, Phys. Rev. Lett. **62** (1989) 1169.

[12] A. Marini, R. Del Sole, G. Onida, Phys. Rev. Lett. **88** (2002) 016403.

[13] F. Aryasetiawan, Phys. Rev. B **46** (1992) 13051.

[14] V. Olevano and L. Reining, Phys. Rev. Lett. **86** (2001) 5962.

[15] W. Ku and A. G. Eguiluz, Phys. Rev. Lett. **82** (1999) 2350.

[16] I. Campillo, A. Rubio and J.M. Pitarke, Phys. Rev. B **59** (1999) 12188.

[17] H. Herenreich and H. R. Phillipp, Phys. Rev. **128** (1962) 1622.

[18] M. A. Cazalilla et al., Phys. Rev. B **61** (2000) 8033.

[19] A. Marini, R. Del Sole, G. Onida, Phys. Rev. B **66** (2002) 115101.

[20] G. Fuster et al., Phys. Rev. B **42** (1990) 7322.

[21] V. P. Zhukov, F. Aryasetiawan, E. V. Chulkov, I. G. de Gurtubay, and P. M. Echenique, Phys. Rev. B **64** (2001) 195122.

[22] E.D. Palik, *Handbook of Optical Constants of Solids*, Academic Press, New York, 1985, p. 835.

[23] S. Albrecht, L. Reining, R. Del Sole, and G. Onida, Phys. Rev. Lett. **80** (1998) 4510; L.X. Benedict, E.L. Shirley, and R.B. Bohn, Phys. Rev. Lett. **80** (1998) 4514; M. Rohlfing and S.G. Louie, Phys. Rev. Lett. **81** (1998) 2312.

[24] See+, for instance, D. M. Roessel and W. C. Walker, J. Opt. Soc. Am. **57** (1967) 835 for the case of LiF.

[25] G. Strinati, Rivista del nuovo cimento, **11** (1988) 1.

[26] F. Bechstedt, K. Tenelsen, B. Adolph, and R. Del Sole, Phys. Rev. Lett. **78** (1997) 1528.

[27] R. Del Sole and R. Girlanda, Phys. Rev. B **54** (1996) 14376.

[28] A. Marini and R. Del Sole Phys. Rev. Lett. **91** (2003) 176402.

[29] G. Strinati, Phys. Rev. Lett. **49** (1982) 1519; Phys. Rev. B **29** (1984) 5718.

For a semiconductor the independent particle polarizability χ_0 is expressed as the Fourier transform

$$\chi_{GG'}(q) = 4 \sum_{v,c} \frac{\langle \psi_v | e^{-i(q+G)\cdot r} | \psi_c \rangle \langle \psi_c | e^{-i(q+G')\cdot r} | \psi_v \rangle}{\epsilon_v - \epsilon_c}. \tag{3.3}$$

ψ_v and ψ_c are the valence and conduction band wavefunctions with associated energies ϵ_v and ϵ_c. Exchange and correlation effects in the system response are incorporated in Eq. (3.2) via the exchange and correlation functional K_{xc} (more precisely the term K_{xc} can be determined from $K_{xc}(r, r') = \frac{dV_{xc}}{d\rho}\Big|_{\rho(r)} \delta(r - r')$, where ρ is the electronic density and V_{xc} is the exchange-correlation potential). The assumption in Eq. (3.2) that $K_{xc} = 0$ amounts to the random phase approximation (RPA), i.e. within the RPA no account is taken of the exchange and correlation on the polarizability of the inhomogeneous electron gas[2] (we note however, that such effects are partially incorporated in the ground state calculations of ψ_v and ψ_c). The polarizability (Eq. (3.2)) is then directly dependent on the independent particle polarizability χ_0. The latter is determined by excitation and de-excitations between the valence and the conduction band (3.3) due to a Born-type perturbation. In most cases this term is evaluated for neutral excitations. Below we propose a way to calculated χ_0 for a finite system (C_{60} and metal clusters) while taking into account the effects of exchange on the polarizability (and on the ground state wavefunctions).

3.3   Excitations in Finite Systems: Role of the Electron–Electron Interaction

In a series of experiments [13, 14] the probability for the removal of one electron from the valence band of the carbon fullerenes (C_{60}) has been measured by bombarding it with electrons. Density functional calculations (DFT) with the local density approximation (without RPA) as well as Hartree–Fock calculations failed to reproduce the excitation probability as a function of the excitation energy [15–17]. Thus an obvious step to remedy (or at least to improve) this situation is to evaluate the removal probability within RPA. However, DFT calculations with RPA for removal probabilities turned out to be computationally very demanding and have not yet been performed. We followed another route by adopting a self-consistent Hartree–Fock procedure, using the so-called variable-phase method to deal with a large number of electrons, and performed an RPA procedure appropriate to electron removal processes. The RPA we use is similar in spirit to the conventional one presented in the preceding section, however we were able to incorporate the effect of exchange on the polarizability which has important consequences on the calculated response of the system, depending on the nature of the probing charge.

---

[2] The macroscopic response of the system ϵ_M is given by the diagonal elements of the inverse dielectric function [8] $\epsilon_{GG}^{-1}(q)$, i.e. by $(1/\left[\epsilon_{GG}^{-1}(q)\right])$. Hence it depends on the non-diagonal elements of ϵ, usually referred to as local fields [9]. The macroscopic dielectric function is determined from the long-wave length limit of ϵ_M (i.e. for $q \to 0$).

3.3.1 Formal Development

Consider a cluster of atoms under the influence of an external time-dependent perturbation $U(\boldsymbol{r}, t)$ that couples to the electronic part of the cluster Hamiltonian. The perturbation may be induced by an impinging electron (that acts as a test charge) or by an electromagnetic pulse. We are interested in the characteristic response of the system quantified in terms of electron removal probabilities from the valence shell.

The dynamic of the electrons in the cluster is governed by the Hamiltonian

$$\widehat{H}(\boldsymbol{r}, t) = \widehat{H}_0(\boldsymbol{r}) + U(\boldsymbol{r}, t), \tag{3.4}$$

where $\widehat{H}_0$ is the self-consistent mean field Hamiltonian in the absence of the perturbation. The solution $\Psi(\boldsymbol{r}, t)$ of the time-dependent Schrödinger equation (atomic units are used throughout)

$$\left[i\partial_t + \widehat{H}(\boldsymbol{r}, t)\right] \Psi(\boldsymbol{r}, t) = 0 \tag{3.5}$$

is written as an antisymmetrized product of single-electron wavefunctions, i.e.

$$\Psi(\boldsymbol{r}, t) = \mathrm{e}^{-iE_0 t} \det \|\psi_i(\boldsymbol{r}, t)\|. \tag{3.6}$$

Here E_0 is the Hartree–Fock energy of the ground state. The value of E_0 is determined as the expectation value (recall that $u \equiv \frac{1}{|\boldsymbol{r} - \boldsymbol{r}'|}$ and the ionic potential experienced by the electrons is denoted by V_{ions})

$$E_0 = \sum_i \langle i| - \frac{\nabla}{2} - V_{\mathrm{ions}} |i\rangle + \frac{1}{2} \sum_{i,k} \langle ik| u |ik - ki\rangle. \tag{3.7}$$

The time dependent single particle orbitals $\psi_i(\boldsymbol{r}, t)$ are then expanded in term of time-independent Hartree–Fock orbitals

$$\psi_i(\boldsymbol{r}, t) = A_i \left[\phi_i(\boldsymbol{r}) + \sum_m C_{mi}(t)\phi_m(\boldsymbol{r})\right]. \tag{3.8}$$

The index m refers to states above Fermi level E_{F} (particle states) whereas i labels the states below E_{F} (hole states). The factor A is a normalization coefficient. From the meaning of the indices i and m one concludes that the expansion coefficients $C_{mi}(t)$ are the probability amplitudes for the creation of the m–i electron–hole pair. The sum in Eq. (3.8) implies a summation over discrete states and an integration over the continuum (particle) states. To obtain a determining equation for the particle–hole excitation amplitudes $C_{mi}(t)$ one inserts Eq. (3.8) into Eq. (3.6) and requires that

$$\langle \Psi(\boldsymbol{r}, t)| \widehat{H} - i\frac{\partial}{\partial t} |\Psi(\boldsymbol{r}, t)\rangle \equiv 0. \tag{3.9}$$

Expanding in $C_{mi}(t) \neq 0$ and accounting for the first non-vanishing terms one obtains the relation

$$
\begin{aligned}
i \sum_{i \leq \varepsilon_{\mathrm{F}} < m} C_{mi}^*(t) \frac{\partial}{\partial t} C_{mi}(t) = \sum_{i \leq \varepsilon_{\mathrm{F}} < m} & \left\{ (\varepsilon_m - \varepsilon_i) |C_{mi}(t)|^2 \right. \\
& + C_{mi}(t) \langle i| U |m\rangle + C_{mi}^*(t) \langle m| U |i\rangle \\
& + \sum_{j \leq \varepsilon_{\mathrm{F}} < k} \left[\frac{1}{2} C_{mi}^*(t) C_{kj}^*(t) \langle mk| u |ij - ji\rangle \right. \\
& + \frac{1}{2} C_{mi}(t) C_{kj}(t) \langle ij| u |mk - km\rangle \\
& \left. \left. + C_{mi}^*(t) C_{kj}(t) \langle mi| u |kj - kj\rangle \right] \right\}.
\end{aligned}
\tag{3.10}
$$

The variation with respect to $C_{mi}^*(t)$ results in the relation

$$
\begin{aligned}
i \frac{\partial}{\partial t} C_{mi}(t) = (\varepsilon_m - \varepsilon_i) C_{mi}(t) + \langle m| U |i\rangle + \\
\sum_{j \leq \varepsilon_{\mathrm{F}} < k} \left[C_{kj}^*(t) \langle mk| u |ij - ij\rangle + C_{kj}(t) \langle mj| u |ik - ki\rangle \right].
\end{aligned}
\tag{3.11}
$$

Solutions of this equation are expressed as

$$
C_{mi}(t) = X_{mi} e^{-i\varepsilon_0 t} + Y_{mi}^* e^{i\varepsilon_0 t},
\tag{3.12}
$$

where ε_0 is the energy imparted by the external perturbation (the incoming projectile). With the ansatz (3.12) we obtain from Eq. (3.11) two coupled equations for the determination of the coefficients X_{mi} and Y_{mi}^*, namely

$$
\begin{aligned}
(\varepsilon_m - \varepsilon_i - \varepsilon_0) X_{mi} + \langle m| U |i\rangle \\
+ \sum_{j \leq \varepsilon_{\mathrm{F}} < k} \left[\langle mj| u |ki - ik\rangle X_{kj} + \langle mk| u |ji - ij\rangle Y_{kj} \right] = 0,
\end{aligned}
\tag{3.13}
$$

$$
\begin{aligned}
(\varepsilon_m - \varepsilon_i + \varepsilon_0) Y_{mi} + \langle i| U |m\rangle \\
+ \sum_{j \leq \varepsilon_{\mathrm{F}} < k} \left[\langle ij| u |km - mk\rangle X_{kj} + \langle ik| u |jm - mj\rangle Y_{kj} \right] = 0.
\end{aligned}
\tag{3.14}
$$

Now we introduce the following definitions for the effective transition amplitudes

$$
\begin{aligned}
-(\varepsilon_m - \varepsilon_i - \varepsilon_0) X_{mi} & =: \quad \langle m| U_{\mathrm{eff}} |i\rangle, \\
-(\varepsilon_m - \varepsilon_i + \varepsilon_0) Y_{mi} & =: \quad \langle i| U_{\mathrm{eff}} |m\rangle.
\end{aligned}
\tag{3.15}
\tag{3.16}
$$

This means that U_{eff} acts as an *effective* external perturbation. Its structure is determined by the naked perturbation and by the particle–hole excitation and de-excitation amplitude. The exact expression for U_{eff} derives from Eqs. (3.13), (3.14), namely

$$\langle m| U_{\text{eff}} |i\rangle = \langle m| U |i\rangle +$$
$$\sum_{j \le \varepsilon_F < k} \left[\frac{\langle k| U_{\text{eff}} |j\rangle \langle mj| u |ki - ik\rangle}{\varepsilon_0 - \varepsilon_k + \varepsilon_j + i\nu} + \frac{\langle j| U_{\text{eff}} |k\rangle \langle mk| u |ji - ij\rangle}{\varepsilon_0 + \varepsilon_k - \varepsilon_j - i\nu} \right]. \quad (3.17)$$

Taking into account that in a linear-response theory $U_{\text{eff}} = \epsilon^{-1} U$ and comparing with Eqs. (3.1)–(3.3) one sees that the approach derived here is equivalent to the RPA, however in the present approach exchange effects on the polarizability are accounted for. We recall that within the HF approximation exchange effects are taken exactly into account but correlation is not accounted for. In contrast, within DFT-LDA exchange and correlation effects are described in an approximate way through the employed exchange and correlation functional. This advantage of the HF comes of the expense of evaluating the expectation value of a large number of non-local potentials (Fock terms) [17, 18]. It turns out that the numerical problems arising from the non-locality of the potentials can be circumvented by utilizing the variable-phase method (VPM). For this purpose we developed a version of VPM applicable to non-local potentials, details are presented in Ref. [19].

3.4   Results and Discussion

A way to test in details (theoretically and experimentally) the features of the matrix elements of the effective interaction U_{eff} is to approach the sample (residing in the initial state ϕ_ν) with a test charge (described by the state vector $|k_0\rangle$, where k_0 is the wave vector of the incoming projectile). The impinging test charge acts on the sample with the perturbation U (usually known in its naked form). This causes the removal of one electron from the bound state (ϕ_ν) into a continuum state, characterized by $|k_1\rangle$, where k_1 is the wave vector of the emitted electron. The final-state wave vector of the projectile is denoted by k_2. According to Eq. (3.17) we write the transition amplitude for this reaction as $T = \langle k_1 k_2 | U_{\text{eff}} | \phi_\nu k_0\rangle$. If electrons are used as a projectile the observable quantity is a spin averaged, differential cross section $W(k_0; k_1, k_2)$ evaluated as the weighted average of the singlet $\propto |T^{(S=0)}|^2$ (vanishing total spin ($S = 0$) of the electron pair) and the triplet $\propto |T^{(S=1)}|^2$ cross sections. Assuming spin-flip processes to be irrelevant, the total cross section $W(\epsilon_0)$ is accordingly given as

$$W(\epsilon_0) = \frac{(2\pi)^4}{k_0} \int d^3 k_1 d^3 k_2 \left\{ \sum_\nu \frac{1}{4} \left| T^{(S=0)}(k_0, \phi_\nu; k_1, k_2) \right|^2 + \right.$$
$$\left. \frac{3}{4} \left| T^{(S=1)}(k_0, \phi_\nu; k_1, k_2) \right|^2 \delta \left(\epsilon_0 + \epsilon_\nu - (k_1^2/2 + k_2^2/2) \right) \right\}. \quad (3.18)$$

If projectiles other than electrons are employed, the exchange part of the cross section vanishes.

Below we present results for the scattering cross section from C_{60} and metal clusters. Using the spherical jellium model [10, 11] we construct the quantum states of the clusters in

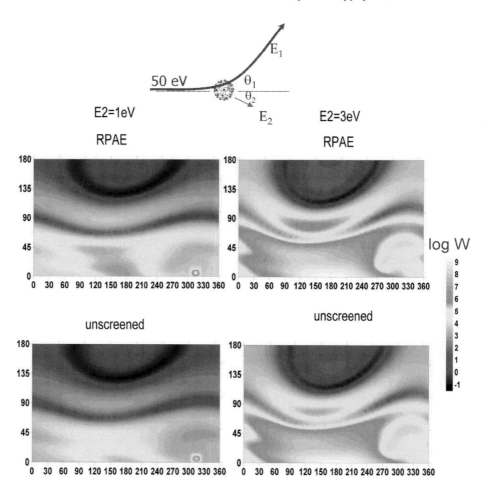

Figure 3.1: The angular dependence of the fully differential cross section for the emission of one electron from C_{60} with 1 eV (left panel) or 3 eV (right panel) following the impact of 50 eV electrons. A schematic of the scattering geometry is depicted. The upper part of the figure shows the RPAE calculation while the lower part shows the calculations without treating screening effects. See also color figure on page 226.

the framework of the Hartree–Fock approximation. The cluster potential (the superposition of atomic potentials) is replaced by a shell confinement. The delocalized valence electrons of carbon atoms are subject to the potential well: $V(r) = V_0$ within the region $R - \Delta < r < R + \Delta$, and $V = 0$ elsewhere. For C_{60} we use $R \approx 6.7\ a_0$ as the radius of the fullerene. The thickness of the shell is $2\Delta \approx 2\ a_0$ (a_0 is the Bohr radius). The height of the well is determined such that the experimental value of the electron affinity of C_{60}^+ and the number of valence electrons are correctly reproduced. A model cluster potential as derived from density functional theory (DFT) within the local density approximation [16] is

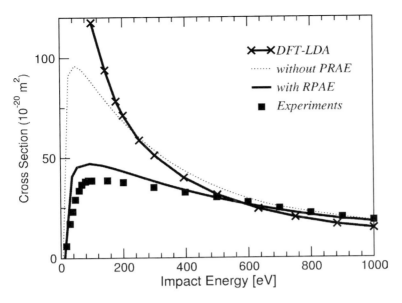

Figure 3.2: The absolute total cross section for the removal of one electron from C_{60} upon the inelastic collision of electrons with the impact energy displayed on the axis. The experimental data (full squares) are taken from Refs. [13,14]. The solid curve with crosses is the result of DFT calculations [15] whereas the dotted curve shows the present theory without RPAE. Theoretical results based on RPAE are shown by the solid curve.

shown to lead to essentially the same conclusions [17]. As demonstrated in Refs. [15, 16], the large extent of C_{60} and the large number of electrons to be considered (240 e^-) results in severe convergence problems in evaluating the transition matrix elements. A way around this problem is to use the non-local variable phase approach [12, 19] for the numerical calculation of the Hartree–Fock states. Upon numerical summation over the states ϕ_ν in Eq. (3.18) one obtains the differential cross sections. To evaluate the total cross section we carry out the six-dimensional integral over the momenta k_1 and k_2 (cf. Eq. (3.18)) using a Monte-Carlo procedure. Figure 3.1 shows the angular dependence of the fully differential cross section $W(k_0; k_1, k_2)$ for C_{60}. The calculations are performed with (upper part in Figure 3.1 and without (lower part in Figure 3.1 inclusion of RPAE (RPA with exchange). In general, the rough structure of the cross section is similar in both cases. The effect of screening is to reduce and broaden some of the peaks. With increasing scattering angle θ_1 of the projectile electron the amount of transferred momentum q to the target increases. Since the response of the system is strongly dependent on the amount of transferred momentum, we observe that the RPA correction are significant only for a certain range of the transferred momentum.

The same applies to the transferred energy dependence of the cross section. This behavior is still observable in the total cross section (Figure 3.2). In the high energy regime there is hardly a difference between the calculations with and without RPA. This is because the time scale for the retarded response of the sample is far off the very short passage time of keV electrons. However at lower energies we observe a strong dependence on screening, in

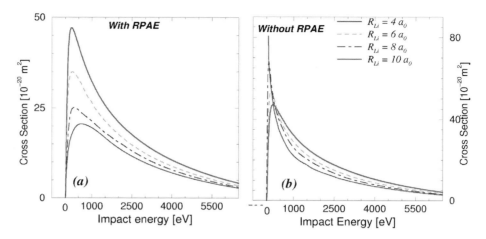

Figure 3.3: The electron-impact ionization cross section for spherical Li clusters with varying radius size R_{Li}. (a) shows the RPAE calculations. (b) shows the results when the particle–hole (de)excitation is neglected.

particular the shape of the cross section is modified. Hence screening of the electron–electron interaction cannot be modelled by a constant suppression of the Coulomb interaction (U/ϵ) since this will lead to a mere energy-independent lowering of the cross section. Figures 3.1 and 3.2 illustrate the ability of the two-electron coincidence technique in mapping the details of the screened electron–electron interaction in small systems, in particular the angular dependence shown in Figure 3.1 is unique to this technique and can not be accessed by other methods such as EELS or photoabsorption spectroscopy.

One of the aims in developing the above theory is to address the system size dependence on the screening of the electron–electron interaction. This is done by considering within the spherical jellium model the electron-removal amplitudes from Li clusters with varying sizes. The cross sections depicted in Figure 3.3 are normalized to the number of electrons in the respective clusters. As evident from Figure 3.3, for small clusters there is hardly any influence of the particle–hole (de)excitations and hence the calculations with and without RPAE are almost identical. On the other hand, with increasing system size the influence of charge density fluctuation (described by the the particle–hole (de)excitations) has a strong influence on the cross section, in particular at low energies.

Contrary to the intuitive expectation, the normalized cross section decreases with increasing cluster size. This is due to an enhanced screening strength, meaning that the effective interaction U_{eff} is reduced. Thus, the scattering region shrinks and consequently the associated cross section decreases.

3.5   Conclusions

This chapter presented a brief overview of how the screening of the electron–electron interaction in finite, nano-size materials can be treated theoretically. The method of choice to map out the details of the electron–electron interaction is the coincident detection of the quantum states of an impinging test charge and a knocked-out electron from the sample. Numerical examples have been presented that illustrate the energy and the angular dependence of the screening effects as well as the dependence on the system size. It should be noted that the calculational examples are the results of the first-order term in a perturbation expansion (cf. Section 3.3.1). In principle the reliability of the results should be assessed by evaluating the contributions of higher-order terms. However the numerical calculations of these terms are far more computationally demanding than RPA and have not yet been explored.

References

[1] A. F. Hebard, M. J. Rosseinsky, R. C. Haddon, D. W. Murphy, S. H. Glarum, T. T. M. Palstra, A. P. Ramirez, A. R. Kortan, Nature **350** (1991) 600.

[2] R. W. Lof, M. A. van Veenendaal, B. Koopmans, H. T. Jonkman, G. A. Sawatzky, Phys. Rev. Lett. **68** (1992) 3924; R. W. Lof, M. A. van Veenendaal, H. T. Jonkman, G. A. Sawatzky, J. Electron Spectrosc. Relat. Phenom. **72** (1995) 83.

[3] D. P. Joubert, J. Phys.: Condens. Mater **5** (1993) 8047.

[4] O. Gunnarsson and G. Zwicknagel, Phys. Rev. Lett. **69** (1992) 957.

[5] P. L. Hansen, P. J. Fallon, W. Kratschmer, Chem. Phys. Lett. **181** (1991) 367.

[6] S. L. Ren, Y. Wang, A. M. Rao, E. McRae, J. M. Holden, T. Hager, KaiAn Wang, Wen-Tse Lee, H. F. Ni, J. Selegue, P. C. Eklund, Appl. Phys. Lett. (1991) 2678.

[7] M. S. Hybertsen and S. G. Louie, Phys. Rev. B **35** (1987) 5585.

[8] R. M. Pick *et al.*, Phys. Rev. B **1** (1970) 910.

[9] A. Baldereschi and E. Tosatti, Phys. Rev. B **17** (1978) 4710.

[10] J. L. Martins, N. Troullier, J. H. Weaver, Chem. Phys. Lett. **180** (1991) 457.

[11] M. Brack, Rev. Mod. Phys. **65** (1993) 677.

[12] F. Calogero, *Variable Phase approach to Potential Scattering*, AP, NY (1967); V. Babikov, *Method of the phase functions in quantum mechanics*, Nauka, Moscow (1971).

[13] S. Matt, B. Dunser, M. Lezius, H. Deutsch, A. K. Becker, B. A. Stamatovic, C. P. Scheier, T. D. Mark, J. Chem. Phys. **105**(1996) 1880.

[14] V. Foltin, M. Foltin, S. Matt, P. Scheier, K. Becker, H. Deutsch, T. D. Mark, Chem. Phys. Lett. **289** (1998) 181.

[15] S. Keller, E. Engel, Chem. Phys. Lett. **299** (1999) 165.

[16] S. Keller, Eur. Phys. J. D **13** (2001) 51.

[17] O. Kidun, J. Berakdar, Phys. Rev. Lett. **87** (2001) 263401.

[18] O. Kidun, J. Berakdar, Surf. Sci. **507-510** (2002) 662.

[19] O. Kidun, N. Fominykh, J. Berakdar, J. Phys. A, **35** (2002) 9413.

4 Electron–Electron Coincidence Studies on Atomic Targets: A Review of (e,2e) and (e,3e) Experiments

Azzedine Lahmam-Bennani

4.1 Introduction

Electron impact single ionization (SI) (e,2e) experiments have been successfully used for the last thirty years to obtain, on the one hand, fundamental information on the dynamics of the SI process and on the details of projectile–target interactions [1], and, on the other hand, direct measurement of the target initial state one-electron wavefunction in momentum space, $|\psi(p)|^2$, via the so-called electron momentum spectroscopy (EMS) [2]. The analogous counterpart for double ionization (DI) is the (e,3e) experiments. In this chapter, I will mainly discuss recent experimental results obtained for the (e,3e) DI, with the aim of trying to emphasize where we are, and what the achievements are in this field. To this purpose, I will use as often as possible comparison with (e,2e) SI results, which are probably more familiar to everyone. Most results discussed here are concerned with the simplest two-electron system, namely helium, however, other rare gases will also be touched upon. For both the (e,2e) and (e,3e) processes, the focus will be on three different aspects: the relationship with the target structure properties or EMS, the behavior of the cross sections in the so-called 'optical limit', and the dynamical aspects of the projectile–target interaction. A review of the theoretical developments in the field as well as other experimental results not discussed here, such as the (e,3-1e) experiments may be found in the recent review by Berakdar et al. [3].

The acronyms (e,2e) and (e,3e) stand for kinematically complete experiments in which energies and momenta of all participating particles are determined in the final state. An incident electron, denoted 0, ionizes the target, and is scattered at an angle θ_a, with energy E_a and momentum k_a. This defines the momentum transfer to the target, $K = k_0 - k_a$. In the (e,2e) case, one electron, denoted b, is ejected from the target in the direction θ_b with energy E_b, while in the (e,3e) case two electrons b and c are ejected. Of course, the electrons are indistinguishable, but for convenience we call them scattered "a" for the fastest one, and ejected "b" and "c" for the slower ones. In the very large majority of studied cases, the target and the residual ion are considered to be in their ground state.

To fully determine the kinematics, one needs to measure all energies and angles, and detect all two (three) final electrons in a double (triple) coincidence studies. Conventionally, this is performed using electrostatic deflection type velocity analyzers [4, 5]. For example, the system used in Orsay, described in detail in Ref. [4], is based on the multi-angle analysis of the two slow ejected electrons in a double toroidal analyzer, using imaging techniques to image the collision plane onto two position sensitive detectors (PSD). The key point is that the

Correlation Spectroscopy of Surfaces, Thin Films, and Nanostructures. Edited by Jamal Berakdar, Jürgen Kirschner
Copyright © 2004 Wiley-VCH Verlag GmbH & Co. KGaA, Weinheim
ISBN: 3-527-40477-5

angular information contained in this plane, and potentially the energy information as well, is preserved upon arrival on the two detectors. An alternative method to the (e,3e) experiments is to detect in coincidence studies the doubly charged ion and two out of the three electrons [6,7]. The system is here based on time of flight techniques: the produced ions are extracted towards an ion detector, whereas the slow ejected electrons are directed along spiraling trajectories onto another PSD located in the opposite direction. The main feature of this system is its large solid angle of collection, a large fraction of 4π. However, the modest momentum resolution achieved makes it necessary to use a supersonic gas jet, and does not allow investigation of heavy targets (presently limited to He and H_2).

Historically speaking, the first (e,2e) experiments were published quasi-simultaneously by Ehrhardt et al. [8] and by Amaldi Jr. et al. [9], with two different goals: in Ehrhardt's case, it was to investigate the ionization process, hence a dynamics aspect. Generally speaking, the angular distributions are here made of two lobes, known as the binary and the recoil lobes [10]. In Amaldi's case, it was to determine the one-electron momentum density (1-EMD) of the target, hence a structure aspect. In the corresponding angular distributions, only the binary lobe remains, the recoil lobe being vanishingly small. Today, more than thirty years later, the distinction between these two aspects has survived, even in the DI case. Of course, these aspects are not fully independent, they are rather – so to speak – nourishing each other.

4.2 Structure Studies

A tremendous amount of work has been done on (e,2e) structure studies by several groups, in particular the groups in Australia, Italy and Canada, see Ref. [11]. The basic idea is that at high enough incoming and outgoing energies, more exactly in the so-called high energy Bethe ridge conditions, the projectile electron acts as a sudden perturbation to the target, interacting solely – in a binary collision – with the target electron which is ejected, the nucleus being only a spectator. That is, all momentum transfer is absorbed by the ejected electron, $K = k_{\rm b}$. Hence, the angular distribution of this electron is a perfect image of the target initial 1-EMD. Or, in other words, the measured triple differential cross section (TDCS) is proportional to $|\psi(p)|^2$. This relationship can be very simply understood from a momentum conservation diagram, see Figure 1(a) of Ref. [12]. Moreover, for ionization of an s electron where the momentum density is maximum at the origin, the angular distribution is also maximum in the K direction, and is symmetrical with respect to K. Whereas for ionization of a p electron where the momentum density has a node at the origin, the angular distribution shows a zero intensity in the K direction. Hence, due to the symmetry with respect to K, the angular distribution has a double-lobe structure [13]. All of this is perfectly verified in the experiments, and has been richly exploited in (e,2e) EMS [11].

What about the (e,3e) case? In (e,3e) DI, a similar diagram is used (Figure 1(c) of Ref. [12]) to illustrate how the initial 2-EMD imposes its signature on the measured (e,3e) distribution. The analog of the Bethe ridge kinematics are now the Bethe sphere kinematics. Here, we require a binary collision between the incident electron and *the pair* of target electrons to be ejected (i.e. with their center-of-mass), the ion once again playing no role, so that all the transferred momentum is absorbed by the electron pair. Without going into detail, there is full analogy with the (e,2e) case: under Bethe sphere conditions, the five fold differen-

(a) He   (b) Ar

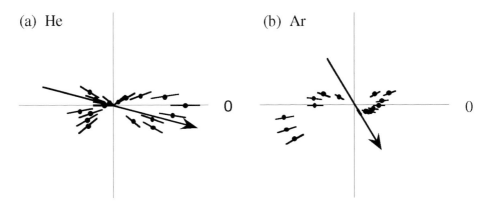

Figure 4.1: Coplanar, relative (e,3e) cross sections for double ionisation, with $E_a = 500$ eV. (a) He($1s^2$), with $E_0 = 601$ eV, $E_b = E_c = 11$ eV and $K = 0.62$ au., (b) Ar($3p^2$), with $E_0 = 561.4$ eV, $E_b = E_c = 9$ eV and $K = 0.81$ au. In both cases, the data are sorted according to the Bethe sphere condition, see Ref. [12]. The 0 indicates the incident direction k_0, and the polar angle is the angle χ of the final state momentum of the ejected electron pair center-of-mass with respect to k_0. The bold arrow indicates the $+K$ direction. Error bars are one standard deviation statistical errors.

tial cross section (5DCS) measured in the (e,3e) experiment *directly images* the target initial 2-EMD, $|\psi(p_1, p_2)|^2$.

However, unlike the (e,2e) case this idea has not yet been fully exploited. The main reason is the small size of the (e,3e) cross sections at large momentum and energy transfers. Up to now, there exist, to my knowledge, only two experimental papers on the subject [12, 14]. To obtain a measurable signal, they both used kinematics which *do not* exactly fulfill the Bethe sphere conditions. Yet, they could observe the 'signature' of the 2-EMD on the measured distributions, as illustrated in Figure 4.1 for DI of He($1s^2$) and Ar($3p^6$): in the He case, the angular distribution of the center-of-mass of the emitted pair of electrons shows a "kind of" binary and recoil lobe distributions, the binary lobe peaking almost in the K direction. Whereas the Ar data show a double lobe structure, with a node, or at least a pronounced minimum, in the K direction. According to the previous understanding of (e,2e) electron momentum spectroscopy and the expected behavior of the 2-EMD, these features can be interpreted as clear signatures of the initial state s- or p-character of the emitted electrons in the ionized orbital.

A word of caution here: The low kinetic energy of the incident and ejected electrons, and the moderate K value prevent a quantitative comparison with 2-EMD calculations, mostly due to second order contributions in the projectile–target interaction. Therefore, future (e,3e) EMS measurements should be performed at higher impact and ejection energies. Recently, Dorn et al. [7] reported COLTRIMS (e,3e) data on He at larger incident energy (2 keV) and momentum transfer (2 au), and moderate ejection energies. They showed that the contributions of the second order effects are less important in the non-coplanar geometry, which thus should be (at least from this point of view) a more appropriate geometry for the (e,3e) EMS. However, as stated above the COLTRIMS technique is presently limited to light targets such as He, and hence awaits further extension to heavier atoms to be meaningfully applied for electron momentum spectroscopy.

4.3 Dynamics Studies

4.3.1 The Optical Limit

It is commonly admitted that in the limit of infinitely high incident energy and vanishing momentum transfer to the target the electron impact ionization cross section asymptotically approaches the photoionization cross section. In the (e,2e) case, this can be written as

$$\lim \Big|_{E_0 \to \infty}^{K \to 0} f_{e2e}^{(3)}(K, E_b, \alpha_b = \theta_{\pm K}) = \frac{3}{4\pi} f_0(E_b) = \frac{3}{8\pi^3 \alpha} \frac{d\sigma_{h\upsilon}}{d\Omega}(E) \tag{4.1}$$

where f_0 is the optical oscillator strength, and $f_{e2e}^{(3)}$ is the electron impact triple differential generalized oscillator strength (TDGOS), given as a function of the (e,2e) TDCS by

$$f_{e2e}^{(3)}(K, E_b, \alpha_b) = \frac{k_0}{2k_a}(E_0 - E_a)\sigma_{e2e}^{(3)}(K, E_b, \theta_b) \tag{4.2}$$

The question I want to address here is how small K needs to be and how large E_0 needs to be for the optical limit to be reached?

The answer for the (e,2e) case is nowadays well established: An example is shown in Figure 4.2 where the TDGOS obtained at the maximum of the binary and the recoil lobes, at

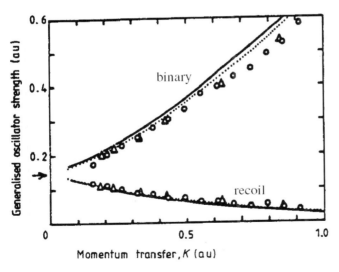

Figure 4.2: Absolute (e,2e) triple differential generalised oscillator strength TDGOS for He, plotted versus momentum transfer K, with $E_0 = 8$ keV (open triangles), and 4 keV (open circles). $E_b = 20$ eV. From Ref. [15]. The arrow on the left axis indicates the optical oscillator strength, f_0.

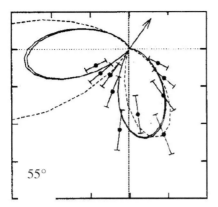 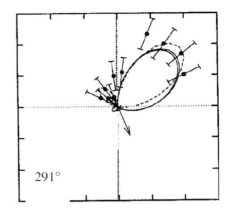

Figure 4.3: Polar plots of the absolute (e,3e) five-fold differential cross sections (5DCS) for He, at $E_0 \sim 5.6$ keV, $K = 0.24$ au and $E_b = E_c = 10$ eV. One ejected electron is detected at a fixed angle, as indicated on the graph by the label and the arrow. The momentum transfer direction is $\theta_K = 319°$. Full line: CCC results (x3.2), dashed line: C4FS results (x0.7). From Ref. [17].

4 and 8 keV incident energy, are plotted versus momentum transfer [15]. It is important to note that the experiments are obtained on an *absolute scale* using a completely independent method based on the Bethe sum rule. Remarkably, the data from both the binary and the recoil branches are found to nicely approach the optical oscillator strength as K goes to zero. Hence the conclusion that at a few keV incident energy and a momentum transfer of the order of 1 or 2 tenths of an au, the optical limit is reached to within some $10 - 15\%$.

The situation is different in the (e,3e) case. Also here, the (e,3e) cross section is expected to converge to the $(\gamma,2e)$ photo-double ionization (PDI) cross section, within a simple and known kinematical factor, via the relationship

$$\lim\Big|_{E_0\to\infty}^{K\to 0} \sigma_{e3e} = 4\frac{k_a}{k_0}\frac{1}{K^2}\frac{c}{4p^2(E_0 - E_a)}\sigma_{\gamma 2e} \tag{4.3}$$

Values of the PDI cross sections are nowadays well established, and the convergent close coupling (CCC) theory [16], for instance, *does* reproduce them correctly. Figure 4.3 shows two examples of (e,3e) angular distributions measured for He at equal energies (10 eV) for the ejected electrons. The data are obtained on an absolute scale, to within an accuracy of about 25% [17]. The impact energy *is* rather large, 5.6 keV, and the momentum transfer *is* rather small, 0.24 au, yet, one observes that the CCC results, shown as a full line, are about a factor of 3 too small. Large overestimate or underestimate factors are also found for calculations with other theoretical models, though the shape is usually well reproduced. This stresses the importance of putting the experiments onto an absolute scale. The observed difference in magnitude is obviously a clear indication that 5 keV is not large enough, and/or 0.2 au is not small enough for the optical limit to be reached. Again, this observation is in contrast with (e,2e) SI where the optical limit is very closely approached under similar kinematics. This

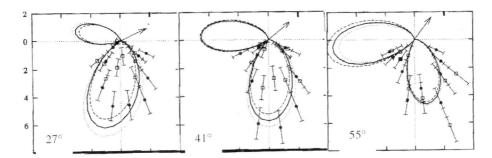

Figure 4.4: Absolute (e,3e) cross sections for He, at $E_0 \sim 5.6$ keV, $K = 0.24$ au and $E_b = E_c = 10$ eV. The data from two fixed angles θ_b and θ_c symmetrical with respect to K are superimposed (full and open dots). The fixed angle θ_b is indicated on the graph by the label and the arrow. The momentum transfer direction is $\theta_K = 319°$. Full line: DPI results, dashed and dotted lines: CCC results for fixed θ_b and θ_c, respectively. From Ref. [17].

emphasizes the fundamental difference between SI and DI, the latter having a much higher ionization threshold, and, most important, being basically induced by the correlation.

There are in fact other ways of looking at the deviations of the (e,3e) data from the optical limit. One possible way concerns the mirror symmetry. In PDI, if one considers two directions for the fixed electron which are symmetrical with respect to the electric field direction ε, then the corresponding angular distributions for the other electron must be symmetrical, that is the mirror image of each other with respect to the ε direction. In electron impact, the momentum transfer direction plays the role of the electric field vector. Therefore, the same mirror symmetry is expected if the optical limit is reached. To investigate this point, samples of the experimental and theoretical results are replotted in Figure 4.4 by superimposing on each graph the data from two angles θ_b and θ_c symmetrical with respect to K (full and open dots). The PI result (black curve) is of course invariant in this symmetry. The two broken curves are the CCC results, which differ only slightly from each other under this symmetry transformation, showing that they are very close to the PDI. Clearly, the experimental results behave differently: almost symmetrical within error bars at some escape angles (e.g. 55°), and appreciably different from each other at some other angles (27° and 41°). We thus conclude that the optical limit is approached differently for different escape angles, for a given incident energy and a given momentum transfer. In other words, the limit depends not only on momentum transfer K and incident energy E_0, but also on the orientation of the vectors k_b and k_c.

4.3.2   Dynamics Studies at Intermediate Energies and Intermediate Momentum Transfer

In this section, I will mostly be comparing experimental results with theoretical models, and I will often refer to the first Born or non-first Born regime. Hence, it is appropriate to briefly define the first Born approximation (FBA). The FBA is a high incident energy approximation

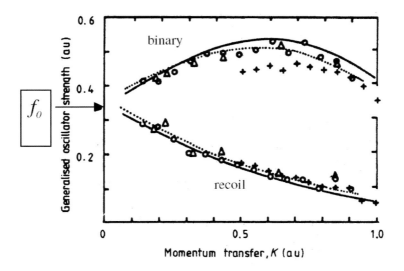

Figure 4.5: Absolute (e,2e) TDGOS for He, plotted versus momentum transfer, K. $E_0 \sim$ 8 keV (open triangles), ~ 4 keV (open circles) and 600 eV (crosses). $E_b = 5$ eV. From Ref. [15].

in which the projectile electron is fast enough to be treated as a small perturbation to the target. Moreover, it is assumed to suffer only one single interaction with the target, hence the designation of the FBA as a first-order model. (For a discussion of interaction mechanisms of the first-order or the non-first order type, see for example Ref. [18]). Therefore, comparison of FB results with experiment will either show good agreement, or else show deviations which could be evidence of collision processes in which the projectile interacts with the target more than once. To establish the grounds for this comparison, we start very briefly with the (e,2e) case.

4.3.2.1   The (e,2e) Case

I will use here two examples which I think nicely summarize the situation in the high and intermediate energy case. In Figure 4.5, the absolute (e,2e) TDGOS measured for He at the maximum of the binary and the recoil lobes is plotted versus momentum transfer, at three incident energies. Within the FBA, this quantity is independent of the incident energy. The full and dashed curves represent two variants of the FB high energy limit of the GOS. Clearly, the data at 8 keV and 4 keV impact energy are almost indistinguishable and have reached the FB limit, both at the binary and recoil lobe whereas the data at 600 eV (crosses) start to seriously deviate from this limit. This is a very general trend as the incident energy is decreased. This trend is nicely summarized in the second example: Shown in Figure 4.6 is the GOS binary-to-recoil ratio at a fixed ejected energy, 5 eV, and a given K value, 0.5 au, plotted versus the inverse of the square root of the incident energy. Again, in the FB model the GOS is independent of incident energy and is represented in this plot by a horizontal line.

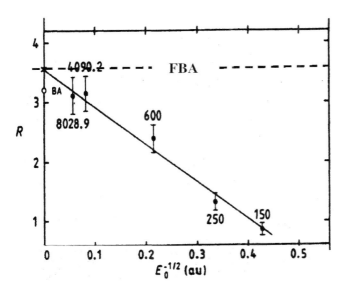

Figure 4.6: Ratio, R of the (e,2e) binary-to-recoil peak intensity for He, plotted versus $E_0^{-1/2}$, at constant ejected energy, $E_b = 5$ eV, constant momentum transfer, $\boldsymbol{K} = 0.5$ au, and at the indicated incident energies. The horizontal line is the FBA prediction. From Ref. [19].

The fact that the data do not fall on this line, but only converge to it as E_0 goes to infinity is a clear indication of the progressive departure from FB as E_0 decreases.

From these two examples, one can quantify this deviation: at 4–8 keV, the FBA is valid to within some 5–10% deviation. At 600 eV, substantial deviations of typically 30% are observed. More sophisticated models, such as second Born treatments [20–22], or inclusion of distorted waves [23,24], are needed and have been successfully introduced. At lower energies, the situation is more complex: I will not discuss it, as there are no (e,3e) DI experiments in this energy regime.

4.3.2.2 The (e,3e) Case

To first understand the structure of the (e,3e) cross section distribution, we consider the theoretical (e,3e) results for DI of He displayed in Figure 4.7, and calculated within a FB- model using a Coulomb wave description of the ejected electrons and a Hylleraas initial state wavefunction for the He atom [25]. The kinematical conditions are those of the experiments in Ref. [17], that is 5.6 keV impact energy. The cross sections are plotted versus the two ejection angles θ_b and θ_c. In all cases discussed in this chapter, the two outgoing electrons have equal energies, 10 eV. Hence, because of their indiscernibility, the data must bear symmetry with respect to the diagonal labeled a in the graph. That is, the information is redundant in this graph: it would have been enough to plot half the surface by cutting along this diagonal. Also, because of the equal outgoing energies, this diagonal is a nodal line, that is a strict zero intensity, due to the Coulomb repulsion which forbids the two electrons from coming out in

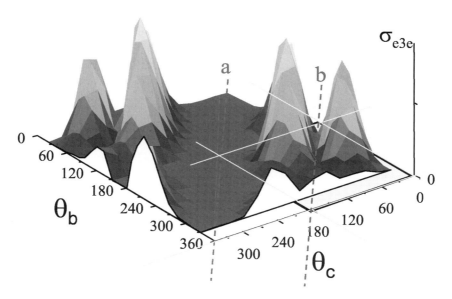

Figure 4.7: The in-plane distribution of the theoretical first Born (e,3e) cross section for DI of He, plotted versus both ejection angles θ_b and θ_c. Kinematical parameters as in [17]: $E_0 = 5.6$ keV, $E_b = E_c = 10$ eV, $\theta_a = 0.45°$ and $K = 0.24$ au. Unpublished results from El Mkhanter [25]. See also color figure on page 227.

the same direction $\theta_b = \theta_c$. The minimum intensity along the line marked b corresponds to the almost forbidden back-to-back emission. Indeed, in PDI, it is well established that for a He target and for equal energy sharing, the angular distribution exhibits a node with a strict zero intensity for two electrons emerging back-to-back, due to the $^1P^o$ symmetry of the electron pair in the final state. In the electron impact case, this node shows up as a minimum of intensity. With these constraints, we see that the intensity distribution is mostly dominated by two structures. Besides, the experimental constraint in the Orsay apparatus to detect both electrons in opposite half planes corresponds to observing only the 'square' at the right corner of the figure, where, fortunately, most of the intensity is. So, in the following, I will mostly be concentrating on this experimentally accessible region.

Figure 4.8 shows this, with the upper row corresponding to theoretical results from the FB-convergent close coupling (CCC) model, at the three energies of the Orsay experiments, that is 5.6 keV, 1.1 keV and 0.6 keV. The theory predicts two 'hills' of intensity (with 'mussel shells' shape), one forward (F) and one backward (B) with respect to the electron beam direction. At the lowest incident energy and largest momentum transfer, the forward hill completely dominates. As the incident energy is increased, and subsequently the momentum transfer is decreased, the backward hill becomes relatively more important, until one reaches the limiting case of PDI (not shown) which corresponds to infinite incident energy and vanishing momentum transfer, where the two hills eventually become identical, and the two 'mussels' become fully symmetric.

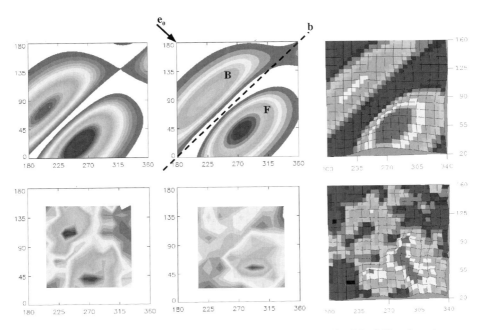

Figure 4.8: The in-plane distribution of the (e,3e) cross section for DI of He, plotted versus both ejection angles θ_b and θ_c. Left column: $E_0 = 5.6$ keV, $E_b = E_c = 10$ eV, $K = 0.24$ au. Middle column: $E_0 = 1.1$ keV, $E_b = E_c = 10$ eV, $K = 0.45$ au. Right column: $E_0 = 0.6$ keV, $E_b = E_c = 11$ eV, $K = 0.61$ au. Upper row: theoretical first Born CCC results, courtesy of A. Kheifets. Lower row: experiments from the Orsay group. The arrow e_0 indicates the incident electron direction, and the dashed line marked b corresponds to the back-to-back emission. See also color figure on page 228.

How does this picture compare to the experiments? The corresponding (e,3e) experimental data are displayed in the lower row of Figure 4.8. Let us first concentrate on the highest energy of 5.6 keV. The comparison there is not too bad: the 'mussel' has lost its regular shape, but this is probably at least partly due to the limited statistics. The nodal line b, which is a reminiscence of the strict zero intensity corresponding to the forbidden back-to-back emission in PDI, is also present in the experiments, but rather as a minimum of intensity than a strict zero. There are two, almost identical, maxima, qualitatively in agreement with theory. The position of the maxima might not be fully the same, but yet, the gross features are there. Reasonable agreement between CCC and experiments is also achieved with the COLTRIMS data of Dorn et al. obtained at 2 keV, see Figure 1 of Ref. [26] where the gross features are correctly reproduced.

However, these observations might be misleading, and deserve some caution: One may be tempted to conclude from these 2D diagrams at 5 and 2 keV that the comparison experiment versus theory is satisfactory, or at least fair, whereas it can be seen from the detailed cuts that *this is not the case*, and that many significant differences are present. For instance, I have already mentioned a factor of 3 difference in the absolute magnitude of the cross sections.

Or else, even if we compare only the shapes of the distributions the agreement is far from being satisfactory at all angles. For example, we have already seen good agreement in shape between theory and experiment at the fixed angles considered in Figure 4.3, but there are some other angles where the agreement is poor, even in shape. Examples may be deduced from Figure 4.4, and can be more clearly seen in Figures 2 and 3 of Ref. [17]. Hence, the necessity of resisting the temptation to use only global 2D-pictures, and making the effort to look at all individual cuts which really contain all the detailed information of these fully differential experiments.

All these deviations are enhanced as the incident energy is decreased to 1.1 keV and 600 eV, Figure 4.8. Being optimistic, one can still find some resemblance between the experimental and theoretical patterns in the fact that the maximum intensity is located in roughly the same angular range. But the 'mussel shape' of the distribution has changed a lot in the experiments:

1. One sees for instance that the forward structure F is still present in both experimental distributions, but the backward one B has not survived, some intensity destruction having occurred, particularly in the 600 eV data.

2. The forward peak in the experimental distribution at 600 eV is stretched along a direction more or less perpendicular to that of the theoretical distribution.

3. This experimental distribution is now made up of various maxima, the largest one at around $60°-280°$ roughly corresponding to the peak position in the FB theoretical results, though being strongly shifted upwards.

4. The back-to-back nodal line, which was clearly seen at 5.6 keV is only vaguely present in Figure 4.8 at lower impact energies.

To visualize these effects in more details, we consider again cuts at fixed θ_b or θ_c, shown in Figure 4.9. I limit the presentation here to cuts taken at $E_0 = 600$ eV in the region around $60°-280°$, that is the values where the first Born results peak. At 50, 60 and 70° fixed angle, the experimental distribution is no longer single-lobed, but also has side lobes or wings. At 280, 290 and 300°, additional lobes are also seen, but the main lobe also exhibits a large shift in its angular position, and is much narrower. Clearly, strong deviations from the FB results are observed, much stronger than discussed above for (e,2e) SI.

The fact that these observations are clearly a manifestation of second or higher order effects, was recently beautifully confirmed by two independent theoretical works, where second Born terms were included: first by Piraux and Dal Cappello [27], who compared their results to the Orsay experiments at 600 eV, and subsequently by Kheifets [28], who implemented second Born contributions in his CCC model and compared the results with the COLTRIMS data at 500 eV and 2 keV. The conclusions from both works are very similar, hence I only show in Figure 4.10 results from Ref. [27]. The FB results of Piraux–Dal Cappello (dashed line) are in fair agreement with FB-CCC results by Kheifets (full line). The results of their model where second Born contributions are included in an approximate way (dotted line) show a large rotation of the peak, clearly towards the experimental lobe. This feature bears an obvious resemblance to previous observations made in (e,2e) SI studies, which strongly suggests

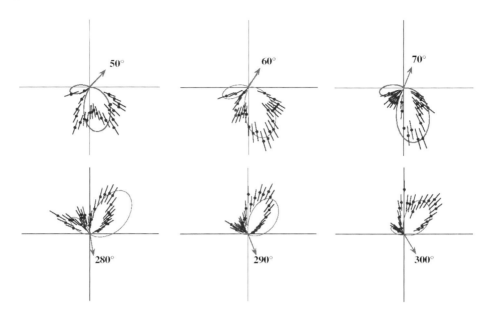

Figure 4.9: Polar plots of the (e,3e) cross section for He, at $E_0 \sim 0.6$ keV, $K = 0.61$ au and $E_b = E_c = 11$ eV. One ejected electron is detected at a fixed angle, as indicated on the graph by the label and the arrow. Full line: FB-CCC results (courtesy of A. Kheifets), dots: experiments (from the Orsay group, partly unpublished) with the associated statistical error bars.

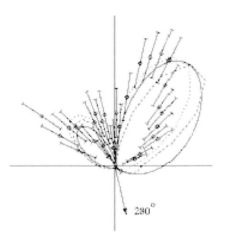

Figure 4.10: Polar plot of the relative (e,3e) cross section for He, at $E_0 = 601$ eV, $K = 0.61$ au and $E_b = E_c = 11$ eV. One ejected electron is detected at a fixed angle, 280° as indicated by the arrow. Dots: experiments. Full line: FB-CCC results (courtesy of A Kheifets), dashed and dotted lines: first and second Born results, respectively, from Ref. [27].

that this effect could be attributed to the presence of large non-first-order contributions in the projectile–target interaction.

Maybe the best proof that such deviations are a manifestation of beyond-first order effects, has been brought by Berakdar [29]. He developed a new approach, based on Green functions (GF) and Fadeev equations, using an iterative procedure where he expresses the interaction

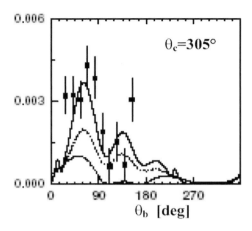

Figure 4.11: The absolute (e,3e) cross section for He, at $E_0 = 5.6$ keV, $K = 0.24$ au and $E_b = E_c = 10$ eV. One ejected electron is detected at a fixed angle, $305°$. Dots: experiments from Ref. [17]. Upper and intermediate curves: GF theoretical results, for electron and positron impact, respectively. Lower curve: FB-CCC results. From Ref. [29].

potential of a system of N-interacting particles in terms of potentials of $(N-1)$ interacting particles, which are further expressed in terms of $(N-2)$ interacting particles, and so on. A sample of the results of this model as compared to the absolute data at 5.6 keV is shown in Figure 4.11. The FB-CCC results (lower curve in the figure) exhibit again the factor of about 3 difference in magnitude with respect to the absolute data. The other two curves are the GF calculations for the electron impact and the positron impact cases, respectively. Two important observations here: first, in the FBA, the results should be insensitive to the charge state of the projectile. Hence, the difference between the e$^-$ and e$^+$ results clearly shows the importance of the non-FB effects, even at this high incident energy of 5.6 keV. Second, there is now a considerably better agreement between experiment and theory, both in shape *and* in magnitude. However, it must be noted that, in a recent paper by Jones and Madison [30], it is argued that the problem of magnitude might rather lie in an improper description of the He-initial state. Using a FB-3C model, with a Pluvinage wavefunction removes the discrepancy with experiments, see Figure 1 of [30]. So, there remains an open question!

As a Summary at this stage, we may say that we observe for DI at ~ 5 keV large non-FB effects. These effects are substantially larger than the $5-10\%$ we saw in (e,2e) SI, (and the about 30% that we can trace in (e,2e) simultaneous ionization+excitation, though I did not discuss this here). However, they mostly affect the magnitude, while the shape of the cross section is less affected. At lower energies, these non-first order effects are obviously much larger for (e,3e) processes than they are for SI at the same impact energy.

Up to this point we have only discussed (e,3e) results for He. The situation is far more complex for heavier targets. I have selected only one example to illustrate this. Figure 4.12 exhibits the angular distribution of the (e,3e) cross section for Ar [31] obtained under kinematical conditions very close to those shown in Figure 4.8 for He at 600 eV. Displayed in Figure 4.12 are both the experimental and the calculated 5DCS distributions based on the FB approximation and using an approximate BBK wavefunction for the final state description. In short, we may note the following points:

1. The directions of the momentum transfer, $+K$ and its opposite, $-K$, the so-called binary and recoil directions in (e,2e) studies, do not play an essential role in defining the structures of the 5DCS distribution.

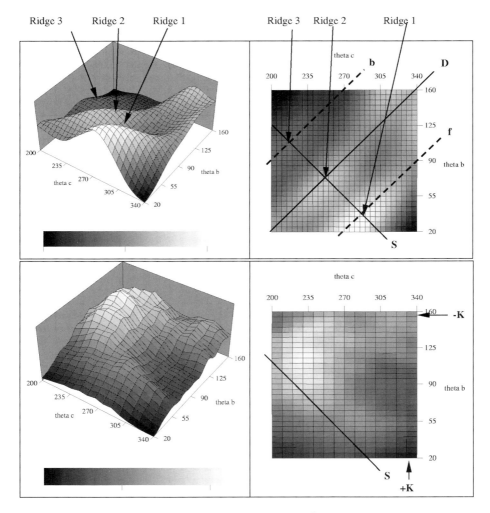

Figure 4.12: Relative (e,3e) cross section for DI of $Ar(3p^{-2})$, versus both ejection angles θ_b and θ_c, at $E_0 = 561$ eV, $E_b = E_c = 9$ eV, $K = 0.4$ a.u. Left column: three-dimensional plot. Right column: contour map projection. Upper row: FB theoretical results. Lower row: experiments. The bold arrows represent the momentum transfer direction, $\pm K$. The line **S** represents the symmetry axis with respect to K, see text. From Ref. [31].

2. As expected from any first order model, the theoretical results exhibit a symmetry axis, labeled S, whose existence corresponds to both electrons being ejected symmetrically with respect to the momentum transfer. In contrast, the experimental data do not show this reflection symmetry.

3. The most striking difference between experimental and theoretical results concerns the overall structure of the cross section distribution. The theoretical results display two

prominent ridges, plus a third, smaller one. The experimental data essentially show one broad, dominant structure, roughly located in an angular range where the theoretical intensity is small.

One might attempt to compare the experimental and theoretical results in more details, by studying the 'individual' cuts through these 3D plots. This has been done [31]. The result is that, apart from some accidental similarities, the theoretical predictions and the experiments display large differences in the number of features, their relative intensities and positions. Clearly, a decisive improvement is needed in the theory.

4.4   Conclusion

In this chapter, a number of (e,3e) results, mostly for DI of He at different impact energies, have been discussed. It has been demonstrated that the problem is much more complex than for (e,2e) SI:

1. The structure EMS studies have become almost a routine in (e,2e), whereas they are still in their infancy for (e,3e).

2. The dynamics aspects seem to be understood for (e,2e), at least at the high and intermediate energies discussed here.

3. For (e,3e), even in the several keV regime, the situation still needs clarification. More experimental data are needed at higher energy, for instance 10 or 15 keV, to check when and how the optical limit is reached?

4. At intermediate energies, strong non-FB effects are obviously present, and still need to be modeled correctly.

5. The results shown for Ar give even stronger evidence of how deep is the gap between theory and experiment. Clearly, a decisive improvement is needed in the theory.

6. On the other hand, it seems also highly desirable to extend the available experimental data to different kinematics and targets to allow a more sensitive comparison with future theoretical developments. To my knowledge, the extension to lower incident energies is under way.

References

[1] Lahmam-Bennani, A. 1991 *J. Phys. B: At. Mol. Opt. Phys.* **24** 2401-2441.

[2] McCarthy, I. E. and Weigold, E. 1991 *Rep. Progr. Phys.* **91** 789.

[3] Berakdar, J., Lahmam-Bennani, A. and Dal Cappello, C. 2003 Phys. Rep. **374** 91-164.

[4] Duguet, A., Lahmam-Bennani, A., Lecas, M. and El Marji, B. 1998 *Rev. Sci. Instrum.* **69** 3524.

[5] Ford, M. J., Doering, J. P., Moore, J. H. and Coplan, M. A. 1995 *Rev. Sci. Instrum.* **66** 3137.

[6] Dorn, A., Moshammer, R., Schröter, C. D., Zouros, T., Schmitt, W., Kollmus, H., Mann, R. and Ullrich, J. 1999 *Phys. Rev. Lett.* **82** 2496.

[7] Dorn, A., Kheifets, A., Schröter, C. D., Najjari, B., Höhr, C., Moshammer, R. and Ullrich, J. 2002 *Phys. Rev. A.* **65** 032709.

[8] Ehrhardt, H., Schulz, M., Tekaat, T. and Willmann, K. 1969 *Phys. Rev. Lett.* **22** 89.

[9] Amaldi Jr., U., Egidi, A., Marconero, R. and Pizzella, G. 1969 *Rev. Sci. Instrum.* **40** 1001.

[10] Ehrhardt, H., Jung, K., Knoth, G. and Schlemmer, P. 1986 *Z. Phys. D* **1** 3.

[11] *Determination of cross-sections and electron momentum profiles of atoms, molecules and condensed matter,* 2002 Special issue of *J. Electron Spectrosc. Relat. Phenom.* **123** 103-409, Eds A. Hitchcock and T. Leung.

[12] Lahham-Bennani, A., Jia, C. C., Duguet, A. and Avaldi, L. 2002 *J. Phys. B: At. Mol. Opt. Phys.* **35** L215-L221.

[13] Daoud, A., Lahham-Bennani, A., Duguet, A., Dal Cappello, C. and Tavard, C. 1985 *J. Phys. B: At. Mol. Phys.* **18** 141-153.

[14] El-Marji, B., Doering, J. P., Moore, J. H. and Coplan, M. A. 1999 *Phys. Rev. Lett.* **83** 1574.

[15] Duguet, A., Chérid, M., Lahham-Bennani, A., Franz, A. and Klar, H. 1987 *J. Phys. B: At. Mol. Phys.* **20** 6145-6156.

[16] Kheifets, A. S. and Bray, I. 1998 *Phys. Rev. A* **58** 4501.

[17] Kheifets, A., Bray, I., Lahham-Bennani, A., Duguet, A. and Taouil, I. 1999 *J. Phys. B: At. Mol. Opt. Phys.* **32** 5047-5065.

[18] McGuire, J. H. 1982 *Phys. Rev. Lett.* **49** 1153; Duguet, A. and Lahham-Bennani, A. 1992 *Z. Phys. D* **23** 383-388.

[19] Ehrhardt, H., Jung, K., Klar, H., Lahham-Bennani, A. and Schlemmer, P. 1987 *J. Phys. B: At. Mol. Phys.* **20** L193-L196.

[20] Byron Jr., F. W., Joachain, C. J. and Piraux, B. 1983 *J. Phys. B: At. Mol. Phys.*, **16** L769.

[21] Marchalant, P. J., Rasch, J., Whelan, C. T., Madison, D. H. and Walters, H. R. J. 1999 *J. Phys. B: At. Mol. Opt. Phys.* **32** L705.

[22] Fang, Y. and Bartschat, K. 2001 *J. Phys. B: At. Mol. Opt. Phys.* **34** L19.

[23] McCarthy, I. E. and Zhang, X. 1989 *J. Phys. B: At. Mol. Opt. Phys.* **22** 2189.

[24] Zhang, X., Whelan, C. T., Walters, H. R. J., Allan, R. J., Bickert, P., Hink, W. and Schönberger, S. 1992 *J. Phys. B: At. Mol. Opt. Phys.* **25** 4325.

[25] El Mkhanter, R. Thèse de Doctorat *Etude des mécanismes de double ionisation de l'hélium par impact électronique* Université de Metz (France) April 1998.

[26] Dorn, A., Kheifets, A., Schröter, C. D., Najjari, B., Höhr, C., Moshammer R and Ullrich, J. 2001 *Phys. Rev. Lett.* **86** 3755.

[27] Lahham-Bennani, A., Duguet, A., Dal Cappello, C., Nebdi, H. and Piraux, B. 2003 *Phys. Rev. A* **67** 010701-4.

[28] Dorn, A., Kheifets, A., Schröter, C. D., Höhr, C., Sakhelashvili, G., Moshammer, R., Lower, J. and Ullrich, J. 2003 *Phys. Rev. A* **68** 012715-4.

[29] Berakdar, J. 2000 *Phys. Rev. Lett.* **85** 4036-39.

[30] Jones, S. and Madison, D. H. 2003 *Phys. Rev. Lett.* **91** 073201.

[31] Jia CC, Lahham-Bennani, A., Dal Cappello, C., Duguet, A. and Avaldi, L. 2003 *J. Phys. B: At. Mol. Opt. Phys.* **36** L17-L24.

5 Studying the Details of the Electron–Electron Interaction in Solids and Surfaces

J. Kirschner, C. Winkler, and J. Berakdar

5.1 Introduction

Recently substantial experimental advances have been achieved in the energy-, spin- and angular-resolved detection of electron pairs that are simultaneously emitted from solid surfaces following excitation by electron or photon beams [1–14]. This process, which is for electron impact also called (e,2e) (one electron in, two electrons out) is visualized schematically in Figure 5.1. The (e,2e) studies from surfaces can be categorized in two broad classes: 1) *the transmission* and 2) *the reflection mode* experiments [1, 2]. The transmission mode experiments are reviewed in Chapter 7. In this kind of experimental arrangement the incoming energetic electron traverses a free-standing thin film. The two emitted electrons are detected on the side of the film that does not contain the incident beam, i.e. both final-state electrons propagate in the forward direction with respect to the incoming beam. A typical reflection mode set-up is shown schematically in Figure 5.1. All the incoming and outgoing vacuum electrons are detected in the same half plane. The information obtainable from the (e,2e) measurements is quite different depending on the experimental arrangement and on the energies and momenta of the vacuum electrons, as discussed below.

5.2 General Considerations

To appreciate the significance of the (e,2e) process it is instructive to inspect the structure of the (e,2e) transition probability $W(\boldsymbol{k}_0, \boldsymbol{k}_1, \boldsymbol{k}_2)$

$$W(\boldsymbol{k}_0, \boldsymbol{k}_1, \boldsymbol{k}_2) \propto \sum_{\substack{i_{\mathrm{occ}} \\ \boldsymbol{k}}} |\langle \Psi_{\boldsymbol{k}_1, \boldsymbol{k}_2} | T_{(\mathrm{e,2e})} | \varphi_{\boldsymbol{k}_0} \chi_{i_{\mathrm{occ}} \boldsymbol{k}} \rangle|^2 \delta(E_1 + E_2 - E_0 - \varepsilon_{\mathrm{i}}). \tag{5.1}$$

Here E_1, E_2 are the energies of the emitted electrons that are described by the final-state wavefunction $\Psi_{\boldsymbol{k}_1, \boldsymbol{k}_2}$ ($\boldsymbol{k}_1, \boldsymbol{k}_2$ are the electrons' wave vectors). The initial electronic states of the target are characterized by the single-particle orbitals $\chi_{i_{\mathrm{occ}} \boldsymbol{k}}$, where $\boldsymbol{k}$ is the Bloch wave vector and the index i_{occ} characterizes all other quantum numbers needed to specify the state of the surface. The incoming electron beam is described by the wavefunction $\varphi_{\boldsymbol{k}_0}$. The operator $T_{(\mathrm{e,2e})}$ induces the transition from the initial state $|\varphi_{\boldsymbol{k}_0} \chi_{i_{\mathrm{occ}} \boldsymbol{k}}\rangle$ into the final state $\langle \Psi_{\boldsymbol{k}_1, \boldsymbol{k}_2}|$ while conserving the total energy.

Correlation Spectroscopy of Surfaces, Thin Films, and Nanostructures. Edited by Jamal Berakdar, Jürgen Kirschner
Copyright © 2004 Wiley-VCH Verlag GmbH & Co. KGaA, Weinheim
ISBN: 3-527-40477-5

Equation (5.1) can be re-written (exactly) as the expectation value of an effective operator M, where

$$W \quad \propto \quad \langle \Psi_{\boldsymbol{k}_1, \boldsymbol{k}_2} | M | \Psi_{\boldsymbol{k}_1, \boldsymbol{k}_2} \rangle, \tag{5.2}$$

$$M \quad = \quad \sum_{\substack{i_{occ} \\ \boldsymbol{k}}} T_{(e,2e)} |\varphi_{\boldsymbol{k}_0} \chi_{i_{occ} \boldsymbol{k}} \rangle \langle \varphi_{\boldsymbol{k}_0} \chi_{i_{occ} \boldsymbol{k}} | \delta(E_1 + E_2 - E_0 - \varepsilon_i) T_{(e,2e)}^\dagger . \tag{5.3}$$

To illustrate the kind of information that can be gained from an (e,2e) experiment we proceed as follows: If in Eq. (5.3) the transition operator $T_{(e,2e)}$ does not depend on the properties of the occupied states[1] we operate within the prior form of scattering theory (i.e. we start from an asymptotic uncorrelated state $|\varphi_{\boldsymbol{k}_0} \chi_{i_{occ} \boldsymbol{k}} \rangle = |\varphi_{\boldsymbol{k}_0} \rangle \otimes |\chi_{i_{occ} \boldsymbol{k}} \rangle$) and then write M as

$$M = T_{(e,2e)} |\varphi_{\boldsymbol{k}_0} \rangle \left\{ \sum_{\substack{i_{occ} \\ \boldsymbol{k}}} |\chi_{i_{occ} \boldsymbol{k}} \rangle \langle \chi_{i_{occ} \boldsymbol{k}} | \delta(E_1 + E_2 - E_0 - \varepsilon_i) \right\} \langle \varphi_{\boldsymbol{k}_0} | T_{(e,2e)}^\dagger$$

$$= \int \mathrm{d}\boldsymbol{q} \mathrm{d}\boldsymbol{q}' \; T_{(e,2e)} |\varphi_{\boldsymbol{k}_0} \boldsymbol{q} \rangle \langle \boldsymbol{q}' \varphi_{\boldsymbol{k}_0} | T_{(e,2e)}^\dagger$$

$$\times \left\{ \sum_{\substack{i_{occ} \\ \boldsymbol{k}}} \chi_{i_{occ} \boldsymbol{k}}(\boldsymbol{q}) \chi_{i_{occ} \boldsymbol{k}}^*(\boldsymbol{q}') \delta(E_1 + E_2 - E_0 - \varepsilon_i) \right\},$$

$$M = -\frac{1}{\pi} \int \mathrm{d}\boldsymbol{q} \mathrm{d}\boldsymbol{q}' \; T_{(e,2e)} |\varphi_{\boldsymbol{k}_0} \boldsymbol{q} \rangle \; \mathrm{Im}\, G(\boldsymbol{q}, \boldsymbol{q}', E) \; \langle \boldsymbol{q}' \varphi_{\boldsymbol{k}_0} | T_{(e,2e)}^\dagger . \tag{5.4}$$

In the second line of this equation we inserted a complete set of plane wave ($\int \mathrm{d}\boldsymbol{q} |\boldsymbol{q}\rangle\langle\boldsymbol{q}|$). The expression in the wavy brackets is the non-local spectral density which is expressible in terms of the imaginary part of the retarded Green's function of the surface $G(\boldsymbol{q}, \boldsymbol{q}', E)$. The latter is evaluated at the energy $E = (E_1 + E_2) - E_0$ that is determined experimentally (all energies are measured with respect to the vacuum level). Inserting Eq. (5.4) into Eq. (5.2) one deduces that

$$W \propto - \int \mathrm{d}\boldsymbol{q} \mathrm{d}\boldsymbol{q}' \; \langle \Psi_{\boldsymbol{k}_1, \boldsymbol{k}_2} | T_{(e,2e)} |\varphi_{\boldsymbol{k}_0} \boldsymbol{q} \rangle \langle \varphi_{\boldsymbol{k}_0} \boldsymbol{q}' | T_{(e,2e)} | \Psi_{\boldsymbol{k}_1, \boldsymbol{k}_2} \rangle \; \mathrm{Im}\, G(\boldsymbol{q}, \boldsymbol{q}', E) \tag{5.5}$$

5.3   Results and Interpretations

The usefulness of the exact mathematical exercise presented above becomes now apparent: Equation (5.5) contains three important physical quantities: 1. The hole non-local spectral density $\mathrm{Im}\, G(\boldsymbol{q}, \boldsymbol{q}', E)$. 2. The electron–electron interaction $T_{(e,2e)}$ and 3. the correlated electron-pair wavefunction $\Psi_{\boldsymbol{k}_1, \boldsymbol{k}_2}$. Choosing the appropriate experimental set-up one can focus on each of these quantities separately. E.g., if we assume that $T_{(e,2e)}$ depend only on

---

[1] This is not always the case, e.g. for certain elements $T_{(e,2e)}$ is spin dependent (for example due to spin–orbit coupling) and hence it depends on the spin-quantum numbers of the electronic states of the surface.

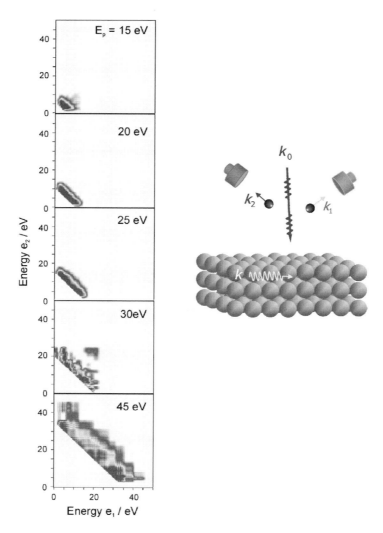

Figure 5.1: A schematic representation of the (e,2e) experimental set-up in the reflection mode geometry is shown on the right hand side. The left part shows a series of energy-distribution spectra of the electron pairs at various impact primary energies shown on the figures ($E_0 = 15$ eV, 20 eV, 25 eV, 30 eV) and $E_0 = 45$ eV. The incoming electron wave vector is perpendicular to the surface (a Cu(001) surface) and the two electrons are emitted symmetrically to the left and to the right of the surface normal at an angle of $40°$. The absolute values of the (e,2e) emission probability are not determined experimentally. See also color figure on page 229.

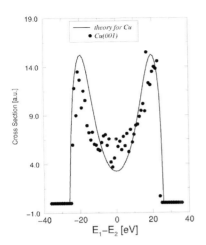
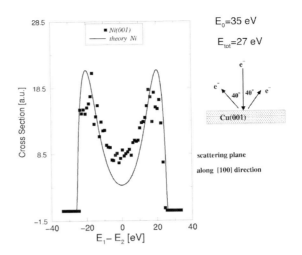

Figure 5.2: The excess energy-sharing of the escaping electron pair with a fixed total pair energy $E_{\text{tot}} = 27$ eV. The incident energy is fixed to $E_0 = 35$ eV. Experimental and theoretical results are shown for a Cu(001) crystal (left panel) and a Ni(001) crystal (right panel). As schematically shown in the figure, the incoming beam is normal to the surface, whereas the two emitted electrons are detected at $40°$ to the left and to the right of the surface normal. The experiments are determined on a relative scale only.

the inter-electron distance and that the vacuum electrons in the initial and the final states occupy plane waves, one finds that $q = q' = k_1 + k_2 - k_0$. Under these assumptions, which are justified at high energies and in the transmission mode, one can study the single-particle Bloch spectral function $\operatorname{Im} G(q, q, E)$, as evident from Eq. (5.5). Such studies are reviewed in Chapter 7. On the other hand, as explained in full detail in Chapter 2, the electron–electron interaction in a many-body system has a rich structure due to the cooperative, correlated behavior of the electrons that results in screening, non-locality and a frequency dependence of $T_{(e,2e)}$. Some experiments to study the properties of $T_{(e,2e)}$ by means of (e,2e) have already been performed [15] (e.g. by tuning the incoming beam wave vector to the plasmon modes of the surface).

The information on the details of interaction between the excited electrons as described by Ψ_{k_1,k_2} can also be singled out, to a large extent, by virtue of the following observation: The energy- and wave-vector conservation laws dictates that $E = E_1 + E_2 - E_0$, $q_\| = k_{1\|} + k_{2\|} - k_{0\|} - GG_\| = q'$. These laws are also expressible in the form

$$\epsilon_{\text{i}} = E_1 + E_2 - E_0, \tag{5.6}$$

$$k_{1\|} + k_{2\|} = k_{0\|} + k + GG_\|. \tag{5.7}$$

Here $GG_\|$ is a surface reciprocal lattice vector. Therefore, we can tune the parameters of the experimental set-up to a certain initially occupied electronic state. E.g., for the (e,2e) coincidence spectra shown in Figure 5.1, all coincidence signals appearing at a constant total energy of the pair $E_{\text{tot}} = E_1 + E_2 = \text{const.}$ originate from the same energy state in the sample (throughout this paper we neglect the effect of inelastic multiple scattering). As deduced from Eq. (5.6), fixing E_0 and varying E_{tot} means varying the initial binding energy ϵ_{i}. This kind of scan can be equally achieved if E_{tot} is fixed and E_0 is varied. Both alternatives are realized in Figure 5.1. To zoom in on the detailed structures present in the density plot shown by Figure 5.1 we choose a certain but fixed $E_{tot} =$ const. I.e. we fix in energy space the initially occupied state of the surface. Then we can vary E_1 and E_2 (while keeping $E_1 + E_2 =$ const.) and study the energy correlation within the pair. This kind of study is performed in Figure 5.2 for Cu(001) and Ni(001). Details of the calculations as well as the technical details of the experiments can be found in various publications (e.g. Refs. [2–7, 9, 11, 12] and references therein). The point which we would like to discuss here in some detail is the typical symmetric shape of the measured spectra with respect to the equal energy point $E_1 = E_2$. This symmetry seems plausible from a geometrical point of view. In reality it is a profound manifestation of the exchange interaction and hence is a signature of one type of electronic correlation. To appreciate this fact we consider in Figure 5.3 a nonsymmetric situation. Figure 5.3(a) shows the calculations neglecting the exchange effects, i.e. when we assume the two electrons to be distinguishable.

Furthermore we consider two situations: Detecting the projectile electron to the right of the incoming beam (at 30°) we obtain the theoretical results shown by the dashed curve. In contrast, if the projectile electron is detected left of the incident beam the dotted curve is obtained. As clear from Figure 5.3(a) the origin of the peaks in the full calculations (including exchange) are certain peaks in the direct and in the exchange scattering probabilities. On the other hand the peaks occurring in the dashed and dotted curves in Figure 5.3(a) are due to the following fact. At moderate and high energies, the quantum electron–electron scattering is most probable when the momentum transfer from one (projectile) electron to the other electron is small. Hence, after being elastically reflected back from the crystal the projectile electron transfers, via electron–electron scattering, little energy to a bound electron and escapes thus with an energy similar to its initial one. In a solid however the bound electron also has a well-defined crystal momentum that varies depending on the initial state. This initial crystal momentum distribution may even dominate the outcome of the encounter with the projectile, as illustrated in Figure 5.3(b–d) where for a fixed impact energy E_0 the kinetic energy E_{tot} is lowered, meaning that the bound electron resides in deeper levels in the conduction band. For all cases Figure 5.3(a–d), it is clear that in the symmetric situation, where both electrons are detected at equal angles (30°) with respect to the incoming beam, the cross section will be symmetric with respect to the equal energy point. In this sense, the symmetric geometry is less favorable for the study of the scattering dynamics.

It should be noted however that the double peak structure of the energy-sharing in Figure 5.2 will change at very low energies to a single flat peak at equal energies. The reason for this is that, quantum mechanical low energy scattering is dominated by s-wave (i.e. isotropic) scattering. On the other hand the electron-pair emission probability has to vanish at $E_1 \to 0$ or $E_2 \to 0$ which in combination with the isotropy of low energy scattering results in one single flat peak at $E_1 = E_2$.

A single peak structure in the energy-sharing distribution can also be obtained at higher energies, but at different emission angles of the electrons, as demonstrated in Figure 5.4 for the case of symmetric electron emission angles with respect to the surface normal. For the emission angles $\theta_1 = \theta_2 = 35°$ we observe the double peak structure, the origin of which has been discussed above. However, when the emission angles are increased the energy region where pair emission may occur shrinks. At $\theta_1 = \theta_2 = 70°$ one single peak at $E_1 = E_2$

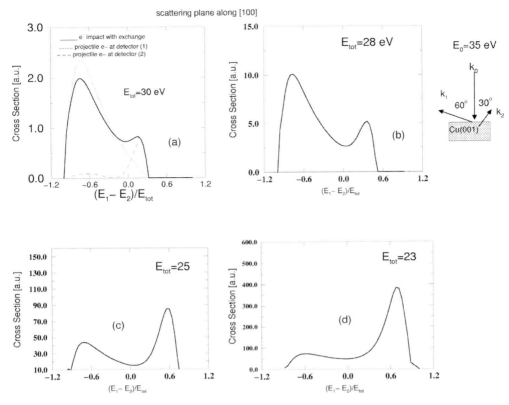

Figure 5.3: The same impact energy as in Figure 5.2 but with varying E_{tot}. In addition one of the detectors (labelled 2) is fixed at an angle of $30°$ to the right of the surface normal whereas the other detector (detector 1) is located at $60°$ to the left of the incoming beam (cf. schematic drawing). The solid curves in (a–d) are the theoretical (e,2e) probability calculated taking account of the exchange effects. In (a) the calculations shown by the dotted and dashed lines are performed while neglecting exchange within the electron pair. The dotted (dashed) line is the outcome of the calculations when the projectile electron is detected with the detector 1 (2). The sample is a Cu(001) single crystal and all vectors (k_0, k_1, k_2) are in one plane (the scattering plane).

is observed, which is actually rather due to the absence of the two peaks at the low energy wings for the case $\theta_1 = \theta_2 = 35°$. The shrinking in Figure 5.4(a),(b) of the energy window for the electron pair emission is dictated by the conservation law for the parallel components of the wave vectors. With increasing emission angles the parallel wave vector component of the fast electron (at the low energy wings) increases. The wave vector conservation law requires however, that the initial state (from which the electron is knocked out) has to contain an equally large parallel-wave vector component. If this is not the case the pair emission is prohibited.

The above discussions and interpretations of the electron-pair spectra did not involve directly the influence of the crystal. However, it is clear that effects such as those shown in

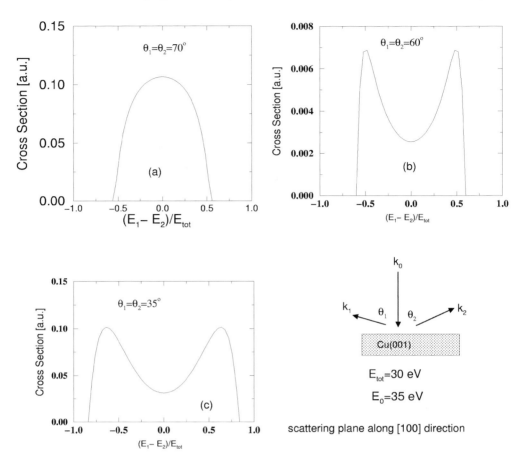

Figure 5.4: The same as in Figure 5.2 however the emission angles of the two electrons are varied as shown on the individual panels. The target is a Cu(001) single crystal.

Figure 5.3 cannot occur in an atomic or a molecular target (for an atomic or a molecular system usually one single discrete bound state is involved in the process and the wave vector is not a good quantum number). In addition, from phase-space consideration one can show [11] that the electron-pair emission probability scales as $k_1 k_2$. Therefore, all the spectra shown above vanish when $E_1 \rightarrow 0$ or $E_2 \rightarrow 0$. This is not the case for an atomic or a molecular target. The reason for this is that, in contrast to metal surfaces, the single electron density of states in an atomic target diverges at threshold as $1/k_{1,2}$. Incidentally this divergence is canceled out by the phase-space factor $k_1 k_2$ and the (e,2e) energy-sharing spectra for atomic or molecular targets are finite when $k_{1,2} \rightarrow 0$.

The ordering of the crystal potential also has a profound influence on the electron-pair energy spectra. A spectrum of a single electron scattered from an ordered multi-center potential shows well-known diffraction patterns [16, 17]. In analogy, two electrons can also be

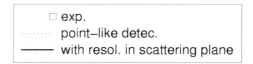

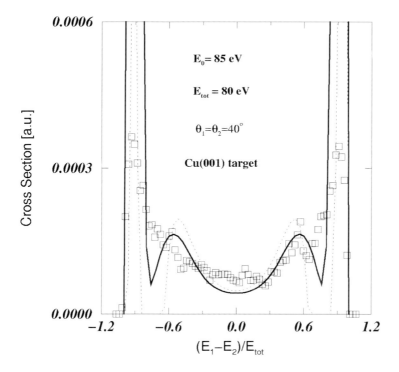

Figure 5.5: The same as in Figure 5.2, however the incident energy is increased to $E_i = 85$ eV and the pair total energy is fixed at $E_{tot} = 80$ eV. The calculations are shown with (solid curve) and without (dotted curve) account for the finite experimental angular resolution (same resolution as in Ref. [3]). Experiments and calculations are performed for a Cu(001) single crystal. The experiments are determined on a relative scale only.

diffracted from the crystal potential, resulting in electron-pair diffraction [3]. As observed in Figure 5.5, when the initial total wave vector of the electron pair $k_{o\parallel} + k \parallel$ is changed by a multiple of a reciprocal lattice vector, diffraction peaks appear. In the experiment these peaks are less pronounced due to the finite experimental resolution (more examples of the pair diffractions are found in Refs. [3, 18]). In this context it should be remarked that the inter-action of the electrons with the crystal potential is usually assumed to be of a single particle nature (but the potential has a multi-center character). Thus, in single-particle terms one can say that one of the electrons is diffracted from the crystal potential and subsequently collides with the other electron. Since the two electrons can exchange momentum, the single particle diffraction peaks are then diffused. On the other hand, other peaks appear due to the conservation of the electron-pair parallel wave vector moduli a reciprocal lattice vector, as expressed by Eq. (5.7). This argument does not rely on which scattering process occurs first. The key point here is that the electron-pair diffraction depends crucially on the momentum transfer within the electron pair and hence it is a direct manifestation of the interaction between the two electrons. This is quite the opposite situation to the diffraction process of a single particle from a crystal potential which is hardly dependent on the electronic correlation.

5.4 Conclusions

The aim of this chapter is to present a small fraction of the rich phenomena that occur when an electron pair is excited from a surface. Here we have chosen to focus on the interpretation of the spin-averaged scattering from clean surfaces. The (e,2e) experiments and theories have evolved however to include the study of ferromagnets, semiconductors, wide-band gap insulators, alloys, clusters and adsorbates. The results of some of these studies are presented in Ref. [19] where further references can be found.

References

[1] S. Iacobucci, L. Marassi, R. Camilloni, S. Nannarone, and G. Stefani, Phys. Rev. B **51** (1995) 10252.

[2] O. M. Artamonov, S. N. Samarin, and J. Kirschner, Appl. Phys. A **65** (1997) 535.

[3] J. Berakdar, S. N. Samarin, R. Herrmann, and J. Kirschner, Phys. Rev. Lett. **81** (1998) 3535.

[4] J. Kirschner, O. M. Artamonov and A. N. Terekhov, Phys. Rev. Lett. **69** (1992) 1711.

[5] J. Kirschner, O. M. Artamonov and S. N. Samarin, Phys. Rev. Lett. **75** (1995) 2424.

[6] O. M. Artamonov, S. N. Samarin and J. Kirschner, Appl. Phys. A **65** (1997) 535.

[7] A. Morozov, J. Berakdar, S. N. Samarin, F. U. Hillebrecht, and J. Kirschner, Phys. Rev. B **65** (2002) 104425.

[8] S. A. Canney, M. Vos, A. S. Kheifets, X. Guo, I. E. McCarthy, and E. Weigold, Surf. Sci. **382**(1-3) (1997) 241.

[9] R. Feder, H. Gollisch, D. Meinert, T. Scheunemann, O. M. Artamonov, S. N. Samarin, and J. Kirschner, Phys. Rev. B **58** (1998) 16418.

[10] A. S. Kheifets, S. Iacobucci, A. Ruocco, R. Camiloni, and G. Stefani, Phys. Rev. B **57** (1998) 7360.

[11] J. Berakdar, M. P. Das, Phys. Rev. A **56** (1997) 1403.

[12] J. Kirschner, O. M. Artamonov, and S. N. Samarin, Phys. Rev. Lett. **75** (1995) 2424; O. M. Artamonov and S. N. Samarin, Tech. Phys. **46** (2001) 1179.

[13] K. A. Kouzakov, J. Berakdar, Phys. Rev. B **66** (2002) 235114.

[14] Z. Fang, R. S. Matthews, S. Utteridge, M. Vos, S. A. Canney, X. Guo, I. E. McCarthy, and E. Weigold, Phys. Rev. B **57** (1998) 12882.

[15] S. Samarin, J. Berakdar, A.Suvorova, O. M. Artamonov, D. K. Waterhouse, J. Kirschner, J.F. Williams, Surf. Sci. **548** (2004) 187-199.

[16] J. B. Pendry, *Low Energy Electron Diffraction* (Academic Press, London, 1974).

[17] M. A. van Hove, W. H. Weinberg, and C.-M. Chan, *Low Energy Electron Diffraction*, in Springer Series in Surface Science (Springer, Berlin 1986).

[18] S. Samarin, J. Berakdar, O. Artamonov, H. Schwabe, and J. Kirschner, Surf. Sci. **470** (2000) 141.

[19] *Many Particle Spectroscopy of Atoms, Molecules, Clusters, and Surfaces*, edited by J. Berakdar and J. Kirschner (Kluwer Academic/Plenum, New York, 2001).

6 Two-Electron Spectroscopy Versus Single-Electron Spectroscopy for Studying Secondary Emission from Surfaces

Sergey Samarin, Oleg Mihailovich Artamonov, Anthony David Sergeant, and James Francis Williams

Experimental results on the (e,2e) reaction on surfaces of a dielectric (LiF film), a metal (W(110) crystal) and a semiconductor (Si(001) crystal) are presented and discussed. A combined analysis of secondary emission spectra together with the (e,2e) spectra of LiF film allows one to establish a link between a "true secondary emission feature" and an energy loss process in the film. A comparison of the (e,2e) spectra of tungsten and silicon shows that the secondary emission mechanisms are different in these materials. The oxygen adsorption strongly modifies the distribution of correlated electron pairs from W(110).

6.1 Introduction

The energy distribution of electrons scattered by a solid surface consists of an elastic (quasi-elastic) maximum and a secondary electron spectrum. The secondary electron spectrum is divided conventionally into two parts: true secondary emission features and energy losses. The origin of the first part is a cascade process of inelastic electron scattering and de-excitation of collective excitations via electron ejection. The energy loss part consists of features due to single-electron excitations (interband transitions) or collective excitations (plasmons, excitons). Techniques based on the analysis of the secondary electron energy distribution (with angular and spin resolution) have become powerful tools for studying the electronic structure of surfaces and electron scattering dynamics.

Inelastic electron scattering from a crystalline surface is usually treated theoretically as a superposition of dipole scattering and impact (or binary) scattering [1, 2]. In the dipolar limit the momentum transfer vanishes and the excitation mechanism is equivalent to electronic excitation by a photon, whereas the impact (binary) limit corresponds to a large momentum transfer of the order of the incident electron momentum. Dipole scattering arises as a result of the Coulomb interaction between an incoming electron and the electric field fluctuations set up in the vacuum outside the target by oscillating surface- and near-surface atoms (charge) [1, 2]. The dipole process occurs most likely outside the target surface, where the electrons undergo small-angle inelastic scattering either preceded or followed by elastic scattering from the surface. However, incident electrons that penetrate the target may be inelastically scattered in the near-surface region by short-range interactions and produce electron–hole pairs. Little is known about the angular distribution of such impact-scattered electrons as they emerge from

Correlation Spectroscopy of Surfaces, Thin Films, and Nanostructures. Edited by Jamal Berakdar, Jürgen Kirschner
Copyright © 2004 Wiley-VCH Verlag GmbH & Co. KGaA, Weinheim
ISBN: 3-527-40477-5

the target. There have been only a few attempts to study the relative contributions of dipole and impact scattering to inelastic electron scattering from surfaces. In one study [3] the electron-spin-labeling technique was used. It was found for Ag and Cu, that have filled 3d-shells, that dipole scattering was dominant, whereas for targets like Mo, Fe, and Co with large densities of unoccupied states the rate for impact scattering was greater than, or comparable to, that for dipole scattering. It was also observed that impact-scattered electrons tend to be concentrated near the specular direction, but not so strongly as dipole-scattered electrons [3]. In the case of highly oriented pyrolitic grafite (HOPG) the scattering mechanism of the electron-energy-loss process in specular reflection geometry was studied [4]. It was shown that an elastic collision always accompanies the inelastic one and two independent channels contribute to the inelastic cross-section, depending on whether the inelastic event precedes or follows the elastic one [4]. The presence of the elastic event associated with the inelastic one implies that in specular reflection geometry the momentum transfer in the inelastic event is minimized and then it can be treated in the dipole approximation [4]. It can be summarized that the fairly complicated phenomenon of electron interaction with a solid surface resulting in a secondary electron emission has been studied extensively but is still not completely understood. One would think that, if both the electrons resulting from an individual electron–electron collision are detected, it would provide insight into the mechanism of secondary emission.

In the recently developed, two-electron-coincidence–spectroscopy-in-reflection mode (referred to as (e,2e) spectroscopy) [5–8], two electrons generated by a single incident electron are detected in coincidence, and the momenta of both electrons are measured. For two electrons generated in a single electron–electron collision, both momentum and energy are conserved and their detection within time intervals of the order of nanoseconds identifies their time correlation and ensures that a minimal number of scattering events is observed [9]. The characteristics of the detected electron pairs contain information on the electron–electron scattering potential and the correlated behavior of electrons inside the solid. This information is related to a near-surface layer of the solid because of the high surface sensitivity of the technique [10]. The spin-polarized version of the (e,2e) spectroscopy has been shown to be a very efficient approach to visualize the spin-dependent scattering dynamics in a ferromagnetic surface [11].

True secondary emission features and energy losses in the secondary emission spectrum are usually considered and discussed separately. In contrast, (e,2e) spectroscopy allows, in some cases, the energy loss structure and the true secondary emission features to be related. Indeed, it may happen that one of the detected electrons of the correlated electron pair is an inelastically scattered primary electron, which has lost part of its energy for plasmon excitation, for example, and the second electron of the pair is an ejected electron resulting from the plasmon decay. If this scattering mechanism is dominant then, in the two-dimensional energy distribution of correlated electron-pairs, a maximum will be observed at electron energies $E_1 = E_e$ and $E_2 = E_p - \hbar\omega_p$, where E_e is the energy of the ejected electron and $\hbar\omega_p$ is the plasmon energy.

In this chapter we present experimental results on the (e,2e) reaction on surfaces of a dielectric, a metal and a semiconductor. The paper is structured as follows. In Section 6.2 the experimental details of low-energy time-of-flight (e,2e) spectroscopy are described. Section 6.3 contains experimental results on the (e,2e) reaction on surfaces of a LiF film, a single crystal of W(110) and a single crystal of Si(100) and their discussion, followed by conclusions in Section 6.4.

6.2   Experimental Details of the Time-of-Flight (e,2e) Spectroscopy in Reflection Mode

6.2.1   Experimental Set-Up

The geometry and electronics of the two-electron coincidence experiment are shown in Figure 6.1. Experiments were carried out in UHV conditions with a base pressure in the $10^{-10}-10^{-11}$ Torr range. The residual magnetic field within the vacuum chamber was reduced to less than 5 mG using Helmholtz coils. A sample was mounted on a movable holder and was cleaned in the vacuum prior to measurements. A Faraday cup (FC) was placed on the axis of the electron gun behind the sample and was used for incident current measurement when the sample was moved off-axis.

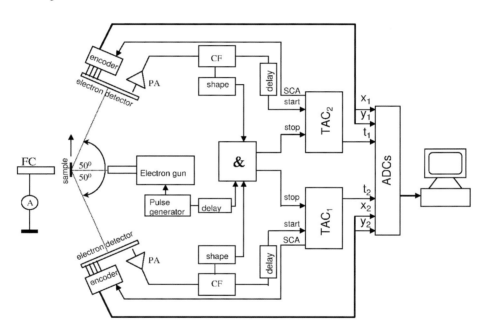

Figure 6.1: Geometry and schematics of the time-of-flight coincidence spectrometer.

A time-of-flight (TOF) technique was used for measuring the energies of both correlated electrons. A reference point on the time scale was obtained by pulsing the incident electron beam. An electron gun produced a pulsed electron beam with a pulse width of less than 1 ns and repetition rate of 4×10^6 Hz.

Position-sensitive detectors with resistive anodes were used for electron detection. Each of the two detectors, 40 mm in diameter, consists of two micro-channel plates in a Chevron arrangement with a resistive anode (Quantar Technology, Model 3394).

6.2.2 Combination of Time-of-Flight Energy Measurements and Coincidence Technique

When an incident electron generates a correlated pair of electrons and they are detected, two pulses from constant fraction discriminators (CF) start two time-to-amplitude converters (TAC). A stop pulse to both TACs comes from a logic unit that delivers a stop pulse only when two delayed and shaped (200 ns width) pulses from the detectors, and a delayed short trigger pulse from the generator, coincide. The width of the two pulses from the two detectors was chosen to be 200 ns to ensure that for a given flight distance of 100 mm all the electrons in the range from E_p (20−40 eV) down to (0.5−0.6) eV are detected. The combination of coincidence technique with the time-of-flight energy analysis provides the energy distribution measurements "in parallel". This means that whatever the electron energy the electron is detected and its energy is determined. This is the advantage of this technique, which enables an overview of two-dimensional energy distribution of correlated electron-pairs. In the present experiment the average incident current was in the 10^{-13} A to 10^{-14} A range, that implies, on average, less than one electron per incident pulse.

Besides the timing pulses, the electron arrival positions on the detectors were measured. The position sensitivity of the detectors allowed: measurement "in parallel" of the angular distribution of the electrons, observation of electron diffraction patterns, estimation of the electron beam size and measurement of the position-dependent flight time that takes into account the difference in flight distances for electrons arriving, for example, at the center or at the edge of the detector. Three analogue pulses from one detector and three pulses from the other representing the electrons' arrival times T_1 and T_2 and positions on the detectors (x_1, y_1, x_2, y_2), are processed by the ADCs and stored in a list-mode file in a computer. The measured distribution of the correlated electron pairs is then a six-dimensional array, which can be projected on any two-dimensional or three-dimensional distributions such as the number of pairs as a function of E_1 and E_2, for example. An MPA-3 multi-parameter acquisition system (FAST ComTec) was used for data collection.

In addition, the above-described two-electron coincidence spectrometer can be used for measuring low energy electron energy loss spectra by switching off the coincidence conditions in the electronic set-up. The characteristics of the spectrometer are described in Ref. [13].

6.2.3 Data Processing

As an example of data processing we present a description of how the six-dimensional array of a measured (e,2e) spectrum is projected onto a two-dimensional energy distribution of the correlated electron-pairs.

As mentioned above, each correlated electron-pair is represented by six numbers: $T_1, x_1, y_1, T_2, x_2, y_2$. Three numbers (T_i, x_i, y_i) define the energy of each electron. To convert these numbers to electron energy we determine first the exact flight distance from the sample to the impact point on the detector and calculate t_0, the time when the primary electron hits the sample (and scattered electrons leave the sample). For an electron detected at time T and at position (x, y) on the detector, the flight time is $t = T - t_0$ and its flight distance is:

$$L = (L_0^2 + x^2 + y^2)^{1/2}, \tag{6.1}$$

where L_0 is the distance from the sample to the center of the detector.

The elastically scattered electrons, with well-defined kinetic energy E_0, are used to calculate t_0. Using the time position T_0 of the elastically scattered electrons, the distance L_0 between the sample and the detector center, and the incident electron energy E_0, we can calculate t_0 using:

$$t_0 = T_0 - L_0 C^{-1} E_0^{-1/2}, \qquad \text{where} \qquad C = (2/m)^{1/2}.$$

Then the energy E of an electron detected with coordinates (x, y) on the detector at time T is calculated as follows:

$$E = L^2 (tC)^{-2} = (L_0^2 + x^2 + y^2) \left[(T - T_0)C + L_0 E_0^{-1/2} \right]^{-2}. \tag{6.2}$$

The use of position sensitive detectors allows measurements of the angular distributions of correlated electron-pairs and, consequently, scanning of component of electron momentum parallel to the surface. For the case of a single crystal sample, this component of electron momentum is conserved. Therefore the measurement of the outgoing electron momenta enables the determination of the parallel component of the valence electron. Figure 6.2 shows how the momenta of the incident electron k_0 and the two outgoing electrons k_1, k_2 are related to the parallel component of the valence electron $q_{||}$. In the experimental geometry shown in Figure 6.1 the accessible range of $q_{||}$ depends on the primary electron energy. For example, for 30 eV primary energy and normal incidence the accessible range of $q_{||}$ is -1.5 Å^{-1} to 1.5 Å^{-1}. This range can be extended by increasing the primary electron energy or by using off-normal incidence.

6.3   Experimental Results and Discussion

The experimental results on low-energy (e,2e) scattering from an insulator (LiF film), a metal (W(110)) and a semiconductor (Si(001)) are presented and discussed in this section.

6.3.1   LiF Film on Si(100)

LiF is a typical insulator with a wide band gap of about 13 eV. It is characterized by a high secondary electron emission yield. Due to the large band gap one can expect a small contribution from a cascade process to the coincidence electron spectrum. Indeed, an electron inside the solid can undergo a collision with a valence electron only if its energy is sufficient to excite the valence electron over the band gap. Given that the electron affinity of LiF is very small, much smaller than the band gap, the electrons with energy below (10 to 11) eV can probably escape from the solid if their momentum points to the vacuum. It is known that the secondary emission spectrum of a LiF film exhibits "true secondary emission" features at about 7 eV and 11 eV [13]. We analyzed one of these features (7 eV) using a combination of single-electron and two-electron spectroscopies.

A LiF film was deposited on a clean Si(001) surface by thermal evaporation from a molybdenum crucible. The thickness of the film was estimated to be 100 Å to 150 Å. Low energy secondary emission spectra from the LiF film at various primary energies were recorded using

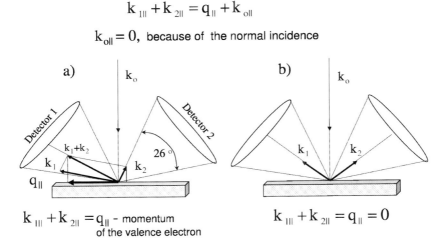

$$\mathbf{k}_{1\|} + \mathbf{k}_{2\|} = \mathbf{q}_{\|} + \mathbf{k}_{o\|}$$

$\mathbf{k}_{o\|} = 0$, because of the normal incidence

$\mathbf{k}_{1\|} + \mathbf{k}_{2\|} = \mathbf{q}_{\|}$ - momentum of the valence electron

$\mathbf{k}_{1\|} + \mathbf{k}_{2\|} = \mathbf{q}_{\|} = 0$

Figure 6.2: Relationship between parallel to the surface components of momenta of the scattered electrons $\boldsymbol{K}_1$ and $\boldsymbol{K}_2$ and the momentum of the bound electron $\boldsymbol{q}$; a) accessible range of $q_\|$ is determined by the combination of $\boldsymbol{K}_1$ and $\boldsymbol{K}_2$ that can be measured; b) combination of $\boldsymbol{K}_1$ and $\boldsymbol{K}_2$ that corresponds to the excitation of the valence electron with $q_\| = 0$, assuming conservation of the components parallel to the surface.

one of the arms of the time-of-flight (e,2e) spectrometer when coincidence conditions were switched off [13]. A set of (e,2e) spectra was measured for the same primary energies. For comparison a secondary emission spectrum for 26.3 eV primary electron energy is shown (Figure 6.3(a)) along with the projection of the two-dimensional energy distribution of correlated electron-pairs on the E_1 axis. In the secondary emission spectrum there are two energy loss features at about 16 eV and 10 eV as well as a prominent true secondary emission feature at about 7.2 eV. The first two correspond to the thresholds of excitation of an excitonic level and the interband transition over the band gap. The maximum at 7.2 eV does not depend on the primary electron energy and therefore is called a "true secondary emission" maximum. The projection of the coincidence spectrum shows the onset at about 13.6 eV and the maximum (7 eV) that coincides with the maximum of the SE spectrum. In the two-dimensional energy distribution (Figure 6.3(b)) there are two maxima, one of which corresponds to the combination of energies $E_1 = (2.6 \pm 0.3)$ eV and $E_2 = (7.2 \pm 0.3)$ eV. The second corresponds to the combination: $E_1 = (7.3 \pm 0.3)$ eV and $E_2 = (2.5 \pm 0.3)$ eV. Systematic measurements of the 2D energy distributions of electron-pairs from LiF film for various primary energies show that above 25 eV incident energy, one electron of the pair is preferentially emitted with $E_1 = (7.2 \pm 0.3)$ eV energy and the second with energy $E_2 = (E_p - 23.3) \pm 0.5$ eV, where E_p is the incident electron energy. This allows the establishment of a link between the true secondary emission feature and the energy loss feature. Indeed, one of the electrons in the maximum of the 2D energy distribution is the one which lost a fixed amount of energy (23.3 eV) for a collective excitation. The second one (7.2 eV) is the electron ejected as a result of the de-excitation process. A detailed discussion of the secondary emission mechanism in a LiF film is the subject of a forthcoming paper [14].

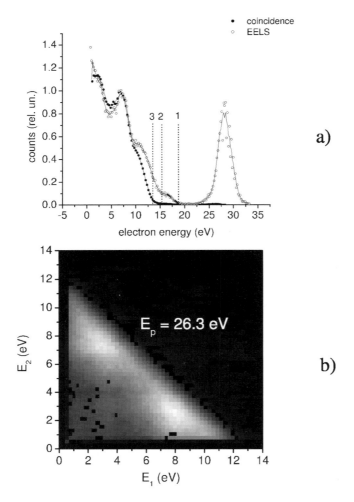

Figure 6.3: a) Electron energy loss spectrum (EELS) (open circles) and projection on E_1 axis (solid circles) of the (e,2e) spectrum of LiF film excited by 26.3 eV primary electrons; b) two-dimensional energy distribution of correlated electron-pairs excited from LiF film by 26.3 eV primary electrons.

The combination of the conventional EELS with the (e,2e) spectroscopy allowed the measurement of the basic parameters of the LiF film: band gap, valence band width, excitonic level position and electron affinity. Such analysis is presented in Ref. [15].

6.3.2   Single Crystal of W(110)

A joint experimental and theoretical study of the (e,2e) reaction on a clean W(001) surface for low energy primary electrons has been carried out recently [16]. In the theoretical treatment, the elastic multiple scattering by the ion cores was taken into account for the primary electron, the valence electron and two detected electrons. The importance of specific elastic events (for example, specular or nonspecular reflection of the incident electron, specular or nonspecular reflection of one or both outgoing electrons) was outlined by additional calculations in which elastic scattering amplitudes were selectively switched off. Generally, elastic reflection was found to be very important in the (e,2e) reaction. Certain features in the (e,2e) spectrum mainly require elastic reflections in the primary electron state, while for others reflection in one of the ejected electron states is required. For valence electron energies a few eV below the Fermi energy, good overall agreement between experiment and theory was found. Since the theory involves only a single direct collision between the projectile and a target electron, this agreement implies that such direct collisions are the dominant origin of the two electrons observed experimentally [16].

Comparison between the calculated (e,2e) spectrum and the density of states in momentum space (*k*-DOS) shows that the main (e,2e) features occur in the regions of high *k*-DOS, whereas in regions of vanishing *k*-DOS the (e,2e) intensity also vanishes. There is, however, no detailed correspondence between (e,2e) spectrum and *k*-DOS. This implies that the incident and ejected electron states play an important role because of their strong elastic multiple scattering by the ion cores. In contrast, at high incident electron energy [17], the primary electron and two outgoing electrons are, to a good approximation, represented by plane waves. Therefore the scattering cross section is proportional to the momentum density of the valence electrons [17].

We present here experimental results on the (e,2e) spectroscopy of W(110). Although the crystallographic face is different from that used in Ref. [16], we believe that the theoretical model and general conclusions are valid here as well.

A clean W(110) surface was obtained by a two-step cleaning procedure [18]. First, the crystal was heated at 1500 K in an oxygen atmosphere at 10^{-7} mbar pressure for several hours to remove carbon impurities. Subsequent flashing to high temperatures (approximately 2500 K) removed the excess oxygen remaining on the surface from the carbon-removal procedure.

The (e,2e) spectra were measured for normal and off-normal incidence at primary energy 25 eV. For off-normal incidence the sample was rotated around the vertical axis by $\pm\alpha$ (the angle is indicated in the figures). Energy sharing distributions in the total energy band of $\Delta E = 2$ eV below the Fermi level, recorded at normal and off-normal incidence (Figure 6.4), indicate that, for off-normal incidence, the energy sharing distribution becomes asymmetric with respect to the zero point where $E_1 = E_2$. If we take the average energy of 19 eV along the total energy band ΔE, then the maximum A in the sharing distribution of Figure 6.4(b) corresponds to the combination of energies $E_1 = 13.2$ eV, $E_2 = 5.8$ eV and, consequently, to momenta $K_1 = 1.86$ Å^{-1} and $K_2 = 1.24$ Å^{-1}. If we assume that these momenta are pointing towards the centers of the detectors, then the summed momentum would run close to the direction of specular reflection. On the basis of this result we can draw the following conclusions: (a) the maximum in the sharing distribution is consistent with the picture in which a correlated electron-pair is generated by the specular beam and the electron with the smaller scattering angle has larger energy; (b) the contribution to the (e,2e) spectrum from bound states with a small component of momentum parallel to the surface is dominant; (c) the kinematics of scattering indicates that the "clean knock out" is the dominant mechanism of the electron-pair generation.

W(110), E_p = 25 eV

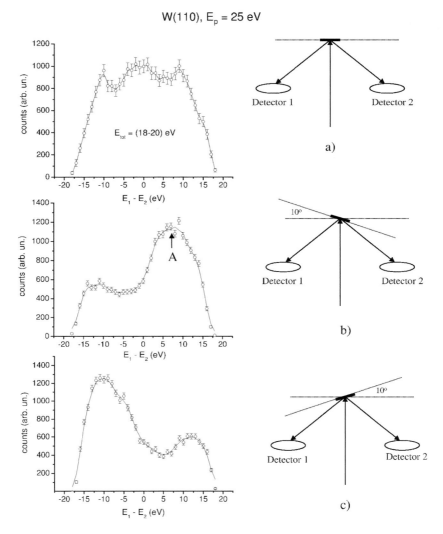

Figure 6.4: Energy sharing distributions of correlated electron-pairs excited from clean W(110) surface by 25 eV primary electrons; a) normal incidence; b) and c) sample tilted by $\pm10°$, as shown in right column.

Using the momentum conservation law for the parallel components of electron momenta: $q_x + k_{0x} = k_{1x} + k_{2x}$, where q is the momentum of the valence electron and k_0, k_1, k_2 are the momenta of the incident and two outgoing electrons, one can plot the number of detected electron-pairs as a function of q_x (Figure 6.5(a)) for two positions of the sample: $\pm12°$ between the sample normal and the bisector between the detectors. We recall here that these curves do not represent directly the momentum density distribution of the valence electrons, as pointed out in Refs. [6, 16]. On the other hand, the momentum density does contribute to the

W(110), E_p=25 eV

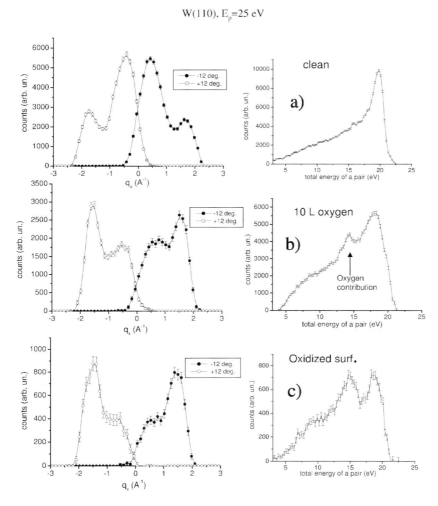

Figure 6.5: "Projections" of (e,2e) spectra on the q_x component of the valence electron (left column) and total energy distributions (right column) of correlated electron-pairs excited by 25 eV primary electron from: a) clean W(110) surface, b) oxygen covered surface, c) oxidized surface. Sample position is indicated on figures ($\pm 12°$).

(e,2e) cross-section with the main (e,2e) features occurring in the regions of high k-DOS [16]. Figure 6.5(a) shows that the contribution from electron states with small $q_x (\pm 0.5$ Å$^{-1})$ is dominant in the case of a clean surface. We tried to change the electronic structure of the surface layer by adsorbing oxygen atoms on the surface and looking at how the distribution of electron-pairs changes as a function of the bound electron momentum. In Figures 6.5(b) and 6.5(c) the corresponding distributions are plotted for oxygen covered (b) and oxidized (c) surfaces. Curve (b) corresponds to the saturated $p(1 \times 1)$ oxygen layer on the W(110) surface that is confirmed by LEED patterns. The presence of oxygen on the surface decreases the contribution of the valence electrons from the near center of the surface Brillouin zone but increases the relative contribution from electron states with larger parallel momentum. It may indicate that the momentum density distribution in the energy band 2 eV below the Fermi level changes upon oxygen adsorption. Indeed, the 2p-oxygen state is "responsible" for the oxygen–metal bonding and can contribute to the (e,2e) spectrum in this momentum range. (We recall that momentum density distribution of atomic oxygen is centered at 1.5 Å^{-1} [19]). In the energy binding spectrum the contribution from an oxygen–adsorbate state shows up at about -6 eV (see right column of Figure 6.5) that is consistent with other measurements [20]. For an oxidized surface (3 min at 10^{-7} Torr pressure of oxygen and 1500 K sample temperature) the relative contribution from bound states with small momenta decreases further (Figure 6.5(c)).

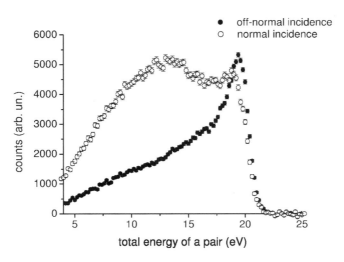

Figure 6.6: Total energy distributions of correlated electron-pairs excited by 25 eV primary electrons from clean W(110) surface at normal (open circles) and 10 degrees off-normal (solid circles) incidence.

The comparison of the total energy distributions of correlated pairs from clean W(110), for normal and off-normal incidence shows, in Figure 6.6 that for off-normal incidence the relative contribution of the pairs with the highest total energy (that corresponds to the excitation of the valence electrons from the vicinity of the Fermi level) increases compared to the normal incidence and forms the maximum at about 19.4 eV. For normal incidence the distribution is broader with a maximum at about $(12-13)$ eV. Given that the time-correlated pairs with small total energy (4 eV below the Fermi edge and lower) result mostly from the multi-step scattering [9], one can conclude that the off-normal incidence increases the probability of single-step electron–electron collisions. The physical reason for this is that at off-normal incidence the momentum transfer, in the electron–electron collision that follows elastic scattering from the ion core, is lower than at normal incidence and consequently leads to the asymmetric sharing distribution.

6.3.3 Single Crystal of Si(001)

Silicon is a typical semiconductor with a band gap of 1.12 eV. A silicon single crystal, with crystallographic plane (001) parallel to the wafer surface, was chosen as a sample. It was chemically cleaned before placing into the vacuum chamber, as described elsewhere [13]. The final stage of the sample cleaning by heating was monitored by LEED and recorded (e,2e) spectra. The evolution of a total energy distribution of correlated electron-pairs during the sample cleaning is shown in Figure 6.7. The maximum at about 15 eV in spectra 1 to 3 (−6.5 eV binding energy) is most likely due to the presence of oxygen (silicon dioxide) that was removed from the surface after heating to 1200°C (curve 4 in Figure 6.7). The maximum in the total energy distribution of the clean sample in the 18 to 22 eV energy range probably contains a contribution from the surface states on Si(100) because it is sensitive to surface contamination.

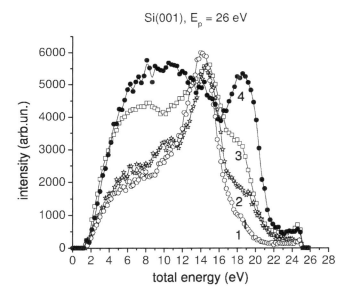

Figure 6.7: Evolution (from curve 1 to 4) of the total energy distribution of correlated electron-pairs excited by 26 eV primary electrons from Si(100) while cleaning the sample.

In contrast to the case of W(110) the energy sharing distribution of pairs excited by 26 eV primary electrons from Si(100) does not show the asymmetry with respect to the zero point when the sample is rotated by 5° or by 25° (Figure 6.8). This means that the scattering mechanism in this case is different from the case of W(110). The reason for this might be a low density of unoccupied electronic states in Si and, consequently, a low joint density of occupied and unoccupied electronic states, which favors dipole scattering [3]. In addition, the screening length in tungsten $\lambda = 0.48$ Å (Thomas–Fermi) is much smaller than in semiconductors ($\lambda = 1-1000$ nm, Debye). This means that the correlation effects are quite different in metals and semiconductors, is in turn can modify the distribution of correlated electron-pairs [21].

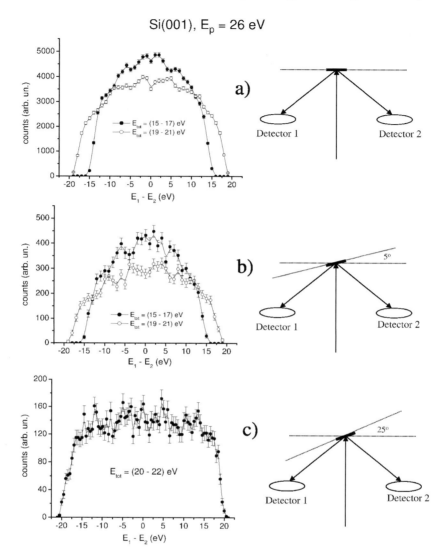

Figure 6.8: Energy sharing distributions of correlated electron-pairs excited by 26 eV primary electron from clean Si(100) surface at: a) normal, b) 5° tilted sample, c) 25° tilted sample.

6.4   Conclusions

We have presented examples of the application of (e,2e) spectroscopy for studying the electron scattering dynamics and the electronic structure of a dielectric, a metal and a semiconductor. We have shown that the true secondary electron features in the secondary emission spectrum of LiF film can be identified using (e,2e) spectroscopy. The link between secondary emission and energy loss process was established.

The combination of (e,2e) spectroscopy and EELS allows the basic band parameters of a LiF film to be measured. A detailed description is presented in Ref. [15].

The kinematics of (e,2e) scattering on W(110) suggest that binary collisions prevail whereas in the case of silicon the scattering mechanism is different.

Oxygen adsorption on a clean W(110) surface changes dramatically the "projection" of the (e,2e) distribution on the q_x component of the valence electron. This change is thought to be partially due to the change in the momentum density distribution in the surface layer upon oxygen adsorption.

References

[1] D.L. Mills, Surf. Sci. **48** (1975)59.

[2] H. Ibach and D.L. Mills, Electron Energy Loss Spectroscopy and Surface Vibrations, Academic, New York, 1982.

[3] H. Hsu, M. Magugumela, B.E. Johnson, F.B. Dunning, and G.K. Walters, Phys. Rev. B **55**, 20 (1997) 13972.

[4] A. Ruocco and M. Milani, S. Nannarone, and G. Stefani, Phys. Rev. B **59**, 20 (1999) 13359.

[5] S. Iacobucci, L. Marassi, R. Camilloni, S. Nannarone, and G. Stefani, Phys. Rev. B **51** (1995) 10252.

[6] A.S. Kheifets, S. Iacobucci, A. Ruocco, R. Camilloni, and G. Stefani, Phys. Rev. B **57** (1998) 7360.

[7] J. Kirschner, O.M. Artamonov, and A.N. Terekhov, Phys. Rev. Lett. **69** (1992) 1711.

[8] J. Kirschner, O.M. Artamonov, and S.N. Samarin, Phys. Rev. Lett. **75** (1995) 2424.

[9] O .M. Artamonov, S.N. Samarin, and J. Kirschner, Appl. Phys. A (Mater. Sci. Proc.), **65**, 6 (1997) 535.

[10] S. Samarin, R. Herrmann, H. Schwabe, and O. Artamonov, J. Electron. Spectrosc. and Relat. Phenom., **96** (1998) 61-67.

[11] S.N. Samarin, J. Berakdar, O. Artamonov, and J. Kirschner, Phys. Rev. Lett., **85** (2000) 1746.

[12] Electron Spectroscopy for Surface Analysis, ed. H. Ibach, Springer-Verlag, Berlin, New York 1977.

[13] S.N. Samarin, O.M. Artamonov, and D.K. Waterhouse, J. Kirschner, A. Morozov, J.F. Williams, Rev. Sci. Instrum. **74**, 3 (2003) 1274.

[14] S. Samarin, J. Berakdar, A. Suvorova, O. M. Artamonov, D. K. Waterhouse, J. Kirschner, and J. F. Williams, Secondary-electron emission mechanism of LiF film by (e,2e) spectroscopy, Surf. Sci., **548** (2004) 187-199.

[15] S. Samarin, O. M. Artamonov, A.D. Sergeant, A.A. Suvorova, and J.F. Williams, Solid State Communs., 129 (2004) 389–393.

[16] R. Feder, H. Gollisch, D. Meinert, T. Scheunemann, O.M. Artamonov, S.N. Samarin, and J. Kirschner, Phys. Rev. B **58** (1998) 16418.

[17] Erich Weigold and Ian E. McCarthy, Electron Momentum Spectroscopy, Kluwer Academic/Plenum, New York 1999.

[18] R. Cortenraada, S.N. Ermolov, V.N. Semenov, A.W. Denier van der Gon, V.G. Gle-bovsky, S.I. Bozhko, and H.H. Brongersma, J. Cryst. Growth **222** (2001) 154.

[19] M. Vos, S.A. Canney, P. Storer, I.E. McCarthy, and E. Weigold, Surf. Sci. **327** (1995) 387.

[20] A.M. Bradshaw, D. Menzel, and M. Steinkilberg, Jpn. J. Appl. Phys. Suppl. 2 (1974) 841.

[21] J. Berakdar, H. Gollisch, and R. Feder, Solid State Commun. **112** (1999) 587.

7 EMS Measurement of the Valence Spectral Function of Silicon – A Test of Many-body Theory

Cameron Bowles, Anatoli S. Kheifets, Vladimir A. Sashin, Maarten Vos, Erich Weigold, and Ferdi Aryasetiawan

7.1 Introduction

The electronic properties of the ground states of semiconductors have been studied both experimentally and theoretically for many years. Thus angle resolved photoelectron spectroscopy (ARPES), especially in combination with tunable synchrotron light sources, has been extensively used to map the dispersion of bands in single crystals. However, in the past the experimental work has concentrated almost exclusively on the measurement of energies, and the theoretical valence-band structure calculations have been tested essentially only in terms of their predictions of eigenvalues. In contrast, relatively little attention has been given to the wavefunction of the electrons, despite the fact that wavefunction information provides a much more sensitive way of testing the theoretical model under investigation. Although the wavefunction cannot be measured directly it is closely related to the spectral momentum density. In the independent particle model it is simply proportional to the modulus square of the one-electron wavefunction,

$$A(\boldsymbol{q}, \omega) = \sum_{\boldsymbol{Gk}} n_{j,\boldsymbol{k}} |\Phi_{j,\boldsymbol{q}}(\boldsymbol{q})|^2 \delta(\boldsymbol{q} - \boldsymbol{k} - \boldsymbol{G}) \delta(\omega - \varepsilon_{j\boldsymbol{k}}), \tag{7.1}$$

where $\Phi(\boldsymbol{q})$ is the momentum-space one-electron wavefunction, j is the band index, $\boldsymbol{k}$ the crystal wave vector, and $n_{j\boldsymbol{k}}$ and $\varepsilon_{j\boldsymbol{k}}$ are the occupation number and energy of the corresponding one-electron state. The reciprocal lattice vector $\boldsymbol{G}$ translates the momentum $\boldsymbol{q}$ to the first Brillouin zone. For an interacting many-electron system the full spectral electron momentum density (SEMD) is given by

$$A(\boldsymbol{q}, \omega) = \frac{1}{\pi} G^-(\boldsymbol{q}, \omega) = \frac{1}{\pi} \frac{1}{[\omega - h - \Sigma(\boldsymbol{q}, \omega)]} \tag{7.2}$$

Here $G^-(\boldsymbol{q}, \omega)$ is the interacting single-hole (retarded) Green's function of the many-electron system, Σ is the self energy and h is the one-electron operator, which includes the kinetic energy and the Coulomb potential from the nuclei and the average of the electron charge cloud density (the Hartree potential). Presuming that the Green's function can be diagonalized on an appropriate basis of momentum-space quasiparticle states $\phi_j(\boldsymbol{q})$ (e.g. orbitals

Correlation Spectroscopy of Surfaces, Thin Films, and Nanostructures. Edited by Jamal Berakdar, Jürgen Kirschner
Copyright © 2004 Wiley-VCH Verlag GmbH & Co. KGaA, Weinheim
ISBN: 3-527-40477-5

in atoms, Bloch waves in crystals) then for a crystal it takes the form [1]

$$A(\boldsymbol{q},\omega) = \sum_{j,\boldsymbol{k},\boldsymbol{G}} |\phi_j(\boldsymbol{q})|^2 \delta_{\boldsymbol{q},(\boldsymbol{k}+\boldsymbol{G})} \frac{1}{\pi} \operatorname{Im} G_j^-(\boldsymbol{k},\omega). \tag{7.3}$$

In the absence of electron–electron interactions the non-interacting Green's function is simply a delta function and Eq. (7.3) reduces to Eq. (7.2). The interacting SEMD contains much more information than simply the band dispersion. The main feature describes the probability of a quasiparticle in band j having momentum $\boldsymbol{k}$ and energy ω. The center of the quasiparticle peak is shifted with respect to the one-electron energy $\varepsilon_{j\boldsymbol{k}}$ and it acquires a width due to the finite quasiparticle lifetime. In addition electron correlation effects can give rise to significant satellite structures.

The full SEMD can be measured by electron momentum spectroscopy (EMS) [1, 2], in which the energies E_0, E_1 and E_2 and momenta $\boldsymbol{k}_0$, $\boldsymbol{k}_1$ and $\boldsymbol{k}_2$ of the incident (subscript 0) and two outgoing electrons (subscripts 1 and 2) in high-energy high-momentum-transfer (e,2e) ionizing collisions are fully determined. From energy and momentum conservation one can determine for each (e,2e) event the binding (or separation) energy of the ejected electron

$$\omega = E_0 - E_1 - E_2, \tag{7.4}$$

and the recoil momentum of the ionized specimen

$$\boldsymbol{q} = \boldsymbol{k}_1 + \boldsymbol{k}_2 - \boldsymbol{k}_0. \tag{7.5}$$

The differential cross section is given by [2]:

$$\sigma(\boldsymbol{k}_0,\boldsymbol{k}_1,\boldsymbol{k}_2,\omega) = (2\pi)^4 k_0^{-1} k_1 k_2 f_{ee} A(\boldsymbol{q},\omega). \tag{7.6}$$

Here f_{ee} is the electron–electron scattering factor, which is constant in the non-coplanar symmetric high-energy (e,2e) kinematics used in the spectrometer at the Australian National University [3, 4]. Thus the (e,2e) cross section is directly proportional to the full interacting SEMD. Since the EMS measurements involve real momenta, the crystal momentum $\boldsymbol{k}$ not appearing in the expression for the cross section, EMS can measure SEMDs for amorphous and polycrystalline materials as well as for single crystals.

The prototype semiconductor silicon has been used as a test-bed to investigate the influence of electron correlations on the SEMD, $A(\boldsymbol{q},\omega)$. Many first-principles calculations have been carried out on bulk silicon, see, e.g., Refs. [5–12]. The majority of these calculations are based on the *GW* approximation to the interacting Green's function [13, 14]. The dispersion of the bulk bands in silicon has been studied with ARPES along high symmetry directions (see, e.g., Refs. [15–17]). However, there has been essentially no experimental data available on the shapes of the quasiparticle peaks (i.e. quasiparticle lifetimes) as a function of momentum, and the satellite density as a function of energy and momentum. These properties of the SEMD arise directly from electron correlation effects and provide stringent tests for approximations to the many-electron problem. First-principles calculations of many physical quantities of interest require the interacting one-particle Green's function as input. It is therefore important to have reliable methods for accurately calculating and testing the real and imaginary parts of the self energy, and hence of $A(\boldsymbol{q},\omega)$.

There are severe difficulties in extracting the full $A(\boldsymbol{q}, \omega)$ from experimental data obtained by other techniques. In ARPES these difficulties include knowing the specifics of the transitions involved, such as the untangling of final-state effects from the initial-state ones, and the strong energy and momentum dependence of the matrix elements [13, 18]. ARPES is also very surface sensitive, which can obscure details of the bulk electronic structure (see e.g. Ref. [17]). In addition there is usually a significant background underlying the photoelectron spectrum and this can hide any continuous satellite contributions. Certain Compton scattering experiments in which the struck electron is detected in coincidence with the scattered photon, so-called $(\gamma, e\gamma)$ experiments, can in principle map out the spectral function [19, 20]. However, the energy resolution is such that it is extremely difficult to resolve even the valence contributions from that of the core electrons [21].

In this chapter we present EMS measurements of the valence spectral momentum density for the prototypical and most studied semiconductor, silicon, and compare the results with calculations based on the independent particle approximation as well as calculations based on many-body approximations to the interacting one-particle Green's function. In Section 7.2 we discuss the experimental technique. The theoretical models are outlined in Section 7.3. The results are discussed and compared with the FP-LMTO calculations and the first-principles many-body calculations in Section 7.4 and, where appropriate, with previous ARPES data. In Section 7.4 the role of diffraction is also discussed and in the last section a brief summary and conclusion is given.

7.2 Experimental Details

An outline of the experimental apparatus (described fully in Refs. [3, 4]) and the coordinate system is shown in Figure 7.1. An electron gun emits a highly collimated 25 keV electron beam, which enters the sample region inside a hemisphere held at $+25$ kV. Thus 50 keV electrons impinge on the target sample along the z-direction, the diameter of the beam being 0.1 mm. The emerging pairs of electrons with energies near 25 keV are decelerated and focused at the entrance of two symmetrically mounted hemispherical electrostatic analyzers. The analyzers detect electrons emerging along sections of a cone defined by $\Theta_s = 44.3°$, which is chosen so that if all three electrons are in the same plane then there is no momentum transferred to the target (i.e. $q = 0$). In the independent particle picture this corresponds to scattering from a stationary electron. If the electrons are not in the same plane (i.e. $\phi_1 \neq \phi_2$) then there is a y-component of momentum with $q_x = q_z = 0$, that is only target electrons with momentum $\boldsymbol{q}$ directed along the y-axis can cause a coincidence event. The electrons are detected by two-dimensional position-sensitive electron detectors, mounted at the exit planes of the analyzers, which measure simultaneously over a range of energies and y-momenta (i.e. range of angles ϕ).

Two pairs of deflection plates mounted inside the high-voltage hemisphere along the sections of the cone can be used to change the effective scattering angle by up to $1°$. In this way [3] one can select nonzero values for the x- and/or the z-component of momentum, the y-component always lying in the range -5 to 5 a.u. (atomic units are used here, 1 a.u. = 1.89 Å^{-1}). This allows one to probe the full three-dimensional momentum space. The two double deflectors are also used to check that the measured momenta correspond to the ex-

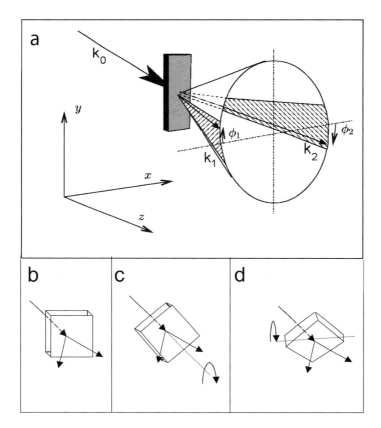

Figure 7.1: Schematics of the experimental arrangements. Incident electrons of momentum k_0 along the z-axis eject an electron from a thin self-supporting Si crystal. The scattered and ejected electrons emerging along the shaded portions of the cone defined by $\Theta = 44.3°$, are detected in coincidence by two energy and angle sensitive analyzers. In the bottom panels the sample is indicated as a block with sides parallel to the $\langle 010 \rangle$ and $\langle 001 \rangle$ symmetry directions. Thus in (b) the spectral momentum density is measured along the $\langle 010 \rangle$ direction. In (c) the density is measured along the $\langle 110 \rangle$ direction, as the crystal has been rotated about the surface normal by $45°$. In (d) the crystal has been tilted by $35.3°$ relative to the position in (c) so that the density is measured along the $\langle 111 \rangle$ direction.

pected ones [3, 22], ensuring that there are no offsets in q_x or q_z as a result of any small possible geometrical misalignments. The orientation of the target specimen can be determined by observing the diffraction pattern of the transmitted electron beam on a phosphorus screen. A specific direction of the thin crystal target is then aligned with the y-axis of the spectrometer so that we measure the energy-resolved momentum density along that direction. The sample orientation can be changed without removing the sample from the vacuum by means of the manipulator mounting arrangement [3].

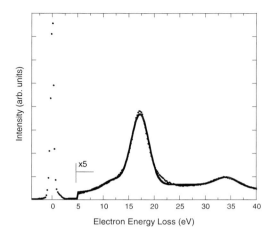

Figure 7.2: The energy loss spectrum observed in the spectrometer for the 20 nm thick Si crystal sample with 25 keV incident electrons. The fitted curve is used to deconvolute the data for inelastic scattering.

The single crystal target has to be an extremely thin self-supporting film. The initial part of the target preparation followed the procedure of Utteridge et al. [23]. First a buried silicon oxide layer was produced by ion implantation in a crystal with $\langle 100 \rangle$ surface normal. A crater was then formed on the back of the crystal by wet chemical etching. The oxide layer serves to stop the etching. It is then removed by a HF dip and the sample is transferred to the vacuum. At this stage the thickness of the thinned part of the crystal is around 200 nm. Low energy (600 eV) argon sputtering is then used to further thin the sample. The thin part of the crystal is completely transparent. The thickness is monitored by observing the color of the transmitted light from an incandescent lamp placed behind the sample. The color changes with thickness due to the interference of the directly transmitted light with that reflected from the front and back silicon-vacuum interfaces. The thinning is stopped when the thickness reaches 20 nm (corresponding to a grey–greenish light). The sample is then transferred to the spectrometer under UHV. A thin amorphous layer could be present on the backside due to the sputtering, whereas the front side (facing the analyzers) is probably hydrogen terminated as a result of the HF dip. The base pressure in the sputtering chamber was of the order of 10^{-9} Torr and in the spectrometer the operating pressure was 2×10^{-10} Torr.

The experimental energy and momentum resolutions were discussed in detail by Vos et al. [1]. The full width at half-maximum (FWHM) energy resolution is 1.0 eV, whereas the FWHM momentum resolutions are estimated to be (0.12, 0.10, 0.10 a.u.) for the q_x, q_y, q_z momentum components respectively.

Even for a very thin target multiple scattering by the incoming or outgoing electrons has to be taken into account. These fast electrons can lose energy by inelastic collisions, such as plasmon excitation, or transfer momentum by elastic collisions (deflection from the nuclei). These scattering events move real coincident (e,2e) events to "wrong" parts of the spectrum, as either the energy or momentum conservation equations (Eqs. (7.4) and (7.5)) are used incorrectly.

For polycrystalline or amorphous solids these multiple-scattering events can be modeled by Monte Carlo simulations [24]. In the case of inelastic scattering the binding energy as inferred from Eq. (7.4) will be too high. Inelastic multiple scattering events can be deconvoluted from the data by measuring an energy loss spectrum for 25 keV electrons passing through the sample. This deconvolution is done without using any free parameters [25]. This approach is used in the present case. Figure 7.2 shows the energy loss spectrum obtained with the present silicon crystal target. For single crystals elastic scattering from the nuclei adds up coherently (diffraction). The change of the incoming or outgoing momenta by diffraction changes the outcome of the measurement by a reciprocal lattice vector. We demonstrate later in this chapter how different measurements can be used to disentangle the diffracted contributions from the primary (non-diffracted) contribution.

7.3 Theory

7.3.1 Independent Particle Approximation

The local density approximation (LDA) of density functional theory (DFT) has long been established as a very useful tool for investigating ground-state properties of bulk semiconductors from first principles [26, 27]. The advantage of DFT for approximate calculations in many-body systems is that one extracts the needed information from a one-body quantity, the electron density $n(\boldsymbol{r})$. Although the one-particle eigenvalues in the theory have no formal justification as quasiparticle energies, in practice they turn out to be surprisingly accurate [28].

We employed here the linear-muffin-tin-orbital (LMTO) method [29] within the framework of DFT. The LMTO method is just one of many computational schemes derived within the framework of the DFT. The great practical advantage of the LMTO method is that only a minimal basis set of energy-independent orbitals (typically $9-16$ per atom) is needed to obtain accurate eigenvalues (band energies). In the present study we implemented a full-potential version (FP-LMTO) of the method [30]. We write the one-electron wavefunction in a crystal in the tight-binding representation as the Bloch sum of the localized MT orbitals:

$$\Psi_{jk}(\boldsymbol{r}) = \sum_{t} e^{i\boldsymbol{k}\cdot\boldsymbol{t}} \sum_{\Lambda} a_{\Lambda}^{jk} \phi_{\Lambda}(\boldsymbol{r} - \boldsymbol{R} - \boldsymbol{t})\,. \tag{7.7}$$

Here $\boldsymbol{k}$ is the crystal momentum, j band index, $\boldsymbol{t}$ lattice (translation) vector and $\boldsymbol{R}$ basis vector. The label Λ defines a MT orbital centered at a given site $\boldsymbol{R}$ and it comprises the site index $\boldsymbol{R}$ and a set of atomic-like quantum numbers. The expansion coefficients a_{Λ}^{jk} are found by solving the eigenvalue problem using the standard variational technique.

Momentum space representation of the wavefunction Ψ_{jk} is given by the Fourier transform of the Bloch function:

$$\Phi_{jk}(\boldsymbol{q}) = \int e^{-i\boldsymbol{q}\cdot\boldsymbol{r}} \Psi_{jk}(\boldsymbol{r})\, \mathrm{d}\boldsymbol{r}. \tag{7.8}$$

Due to the periodic nature of the charge density the only non-zero contributions to $\Phi_{jk}(\boldsymbol{q})$ occur when $\boldsymbol{q} = \boldsymbol{k} + \boldsymbol{G}$ with $\boldsymbol{G}$ a reciprocal lattice vector.

$$\Phi_{jk}(\boldsymbol{q}) = \sum_{\boldsymbol{G}} c_{\boldsymbol{G},k}^{j} \delta(\boldsymbol{q} - \boldsymbol{k} - \boldsymbol{G}). \tag{7.9}$$

The contributions $c_{G,k}^j$, the Bloch wave amplitudes, are expressed through the (Fourier) integrals:

$$
\begin{aligned}
c_{G,k}^j &= \Omega^{-1} \int e^{-i(k+G)\cdot r} \Psi_{jk}(r) \, dr \\
&= \Omega^{-1} \sum_\Lambda \left[a_\Lambda^{jk} e^{-i(k+G)\cdot R} \int e^{-i(k+G)\cdot r} \phi_\Lambda(r) \, dr \right]
\end{aligned}
\tag{7.10}
$$

Here it is assumed that the wavefunction Ψ_{jk} is normalized in the unit cell of the volume Ω. The limits of the three-dimensional integration indicates symbolically the whole coordinate space. The EMD in the occupied part of the band j is proportional to the modulus squared Bloch amplitudes:

$$
\rho_j(q) = \frac{\Omega^2}{(2\pi)^3} \sum_G n_{jk} \left| \Phi_j(k+G) \right|^2 \delta_{q,k+G}
\tag{7.11}
$$

where n_{jk} is the occupation number. The EMD (7.11) is normalized to the total number of valence electrons per unit cell:

$$
2 \sum_j \int dq \, \rho_j(q) = n_e \, .
\tag{7.12}
$$

7.3.2 Electron Correlation Models

The hole Green's function entering Eq. (7.3) can be calculated by the many-body perturbation theory (MBPT) expansion on the Bloch wave basis (7.7). Taking the first non-vanishing term in the MBPT leads to the so-called *GW* approximation [31, 32]. In this acronym *G* stands for the Green's function and *W* denotes the screened Coulomb interaction. The *GW* approximation is known to give accurate quasiparticle energies [14]. However, its description of satellite structures is not satisfactory. In alkali metals, for example, photoemission spectra show the presence of multiple plasmon satellites whereas the *GW* approximation yields only one at too large an energy. This shortcoming of the *GW* approximation has been resolved by introducing vertex corrections in the form of the cumulant expansion to the Green's function [33–35]. This allowed the inclusion of multiple plasmon creation. As a result the calculated peak positions of the plasmon satellites were found to be in much better agreement with the experiment than those predicted by the *GW* scheme itself [36–38].

Formally, the cumulant expansion for the one-hole Green's function can be derived as follows. We choose the time representation for the Green's function, drop the band index j for brevity, and write it as

$$
G(k, t < 0) = i\theta(-t) e^{-i\omega_k t + C^{\mathrm{h}}(k,t)},
\tag{7.13}
$$

where ω_k is the one-electron energy and $C^{\mathrm{h}}(k, t)$ is defined to be the cumulant. Expanding the exponential in powers of the cumulant we get

$$
G(k, t) = G_0(k, t) \left[1 + C^{\mathrm{h}}(k, t) + \frac{1}{2} \left[C^{\mathrm{h}}(k, t) \right]^2 + \dots \right],
\tag{7.14}
$$

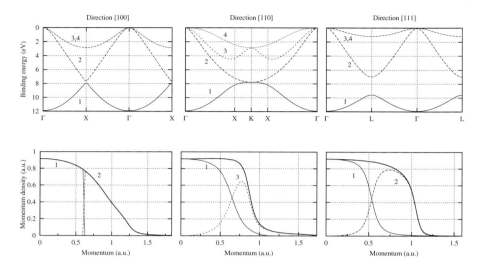

Figure 7.3: FP-LMTO calculations [30] of the dispersion (top panel) and momentum density (bottom panel) of Si for different high symmetry crystallographic directions. The total momentum density is split into the densities of the individual bands, as indicated.

where $G_0(\boldsymbol{k}, t) = i\theta(-t)\exp(-i\omega_k t)$. In terms of the self-energy Σ, the Green function for the hole can be expanded as

$$G = G_0 + G_0 \Sigma G_0 + G_0 \Sigma G_0 \Sigma G_0 + \ldots . \tag{7.15}$$

To lowest order in screened interaction W, the cumulant is obtained by equating

$$G_0 C^{\mathrm{h}} = G_0 \Sigma G_0, \tag{7.16}$$

where $\Sigma = \Sigma_{GW} = iG_0 W$. The first-order cumulant is therefore

$$C^{\mathrm{h}}(\boldsymbol{k}, t) = i \int_t^\infty \mathrm{d}t' \int_{t'}^\infty \mathrm{d}\tau e^{i\omega_k \tau} \Sigma(\boldsymbol{k}, t) . \tag{7.17}$$

This is then put back into Eq. (7.13) yielding multiple plasmon satellites. The energy-momentum representation of the Green's function can be restored by the time Fourier transform.

7.4   Results and Discussions

7.4.1   Band Structure

We first discuss the dispersion $\varepsilon_{j\boldsymbol{k}}(\boldsymbol{q})$, i.e. the dependence of the energy of the Bloch function $\Phi_{j\boldsymbol{k}}(\boldsymbol{q})$ of band j on its crystal momentum $\boldsymbol{k} = \boldsymbol{q} + \boldsymbol{G}$. The band with the largest binding

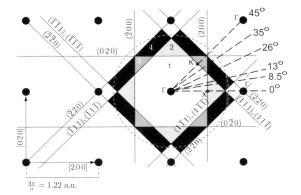

$\frac{4\pi}{a} = 1.22$ a.u.

Figure 7.4: The cut through the reciprocal lattice of silicon along the $q_z = 0$ plane with the first four Brillouin zones labeled. The Brillouin zone boundaries are labeled by the indices of the reciprocal lattice they bisect. The dashed circle is the Fermi sphere for a free electron solid with the same electron density as silicon. Different measurements through the $q = 0$ point are indicated by the dashed lines.

energy is labeled 1, the next 2 etc. In momentum space the Bloch function with crystal momentum $\boldsymbol{k}$ is non-zero only at momentum values $\boldsymbol{k} + \boldsymbol{G}$ with amplitude $c_{\boldsymbol{G},\boldsymbol{k}}^{j}$ (see Eq. (7.9)).

The band structure and momentum densities obtained by Kheifets et al. [30] in a FP-LMTO calculation (based on the DFT-LDA) are shown in Figure 7.3, the bands being labeled as discussed above. The bands are periodic in $\boldsymbol{q}$ space, with band 1 having a maximum in the binding energy at the Γ points. However, the only Γ point with significant momentum density in band 1 is the one corresponding to zero momentum (see lower panel in Figure 7.3). Thus the function with the lowest energy is a Bloch function with $k = 0$, $c_{(0,0,0),\boldsymbol{k}}^{1} \simeq 1$, and the other $c_{\boldsymbol{G},\boldsymbol{k}}^{1} \simeq 0$.

The cut of the first four Brillouin zones of silicon along the $q_z = 0$ plane is shown in Figure 7.4. For a free electron solid the wavefunctions are plane waves and in the ground state the occupied states are within the Fermi sphere with radius k_f i.e. $|k| < k_f$. The intersection of this sphere with the $q_z = 0$ plane is indicated by the dashed circle in Figure 7.4. The lattice potential of silicon can be viewed as a perturbation on the free electron picture, so that the wavefunctions are Bloch functions with more than one $|c_{\boldsymbol{G}k}^{j}|^2 > 0$. The semiconductor silicon has eight valence electrons per unit cell, and hence the first four bands are fully occupied with one spin up and one spin down electron per band. We will now discuss the results of the measurement of silicon and emphasize that the electron density for band j is at its maximum in Brillouin zone j.

The sample was a thin ($\simeq 20$ nm) single silicon crystal with $\langle 001 \rangle$ surface normal, which is first aligned with the z direction (i.e. aligned with $\boldsymbol{k}_0$, see Figure 7.1 (a)). Rotating around the surface normal, measurements were taken with the sample $\langle 100 \rangle$ direction aligned along the y-axis, then the $\langle 110 \rangle$ direction and four intermediate directions were aligned along the y-axis as shown in Figure 7.4. In all these cases the potentials on the sets of deflector plates were set to ensure that the measurements passed through zero momentum (corresponding to $\Gamma_{(0,0,0)}$).

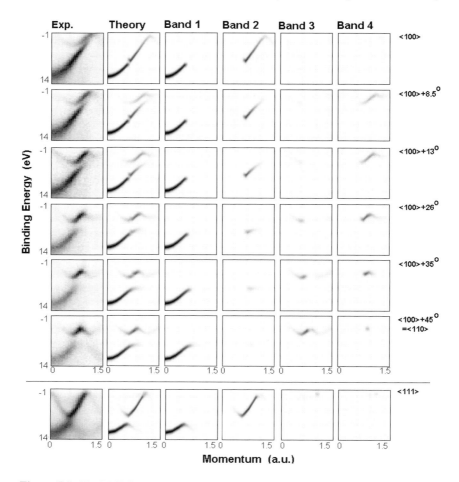

Figure 7.5: The LMTO calculated spectral momentum densities (total plus separate band contributions) along the directions shown in Figure 7.4 compared with the measured results. The calculations have been broadened by the experimental energy resolution. The bottom panel shows the results along the ⟨111⟩ direction.

The experimentally observed density distributions, together with the results of the FP-LMTO calculation, are shown in Figure 7.5 for the six directions measured through $\Gamma_{(0,0,0)}$. The calculations were broadened with the experimental 1 eV energy resolution and split into the four occupied bands.

For the measurement of momenta directed along the ⟨100⟩ direction (0° in Figure 7.4) the theory predicts bands 1 and 2 occupied. In the first Brillouin zone band 1 is occupied, changing abruptly to band 2 at 0.61 a.u. (see also Figure 7.4 and the left panel of Figure 7.3). There is no band gap in the dispersion on crossing the first Brillouin zone. This is due to additional symmetry of the diamond lattice (see, e.g., Ref. [39]). After leaving the second

Brillouin zone the calculated density drops only gradually to zero. The measured density (left panel of Figure 7.5) has the same behavior , but also shows an additional branch at smaller binding energies, which merges with the main feature at 1.2 a.u. This additional branch can also be seen in the calculated band structure for the case where the crystal has been rotated by 8.5°, and comes from band 4. From the shape of the Brillouin zone shown in Figure 7.4 it is clear that the measurement along the $\langle 100 \rangle$ direction just misses Brillouin zone 4. Due to finite momentum resolution it is obvious that the measurement will pick up contributions from this zone, giving rise to the extra branch in the observed intensity.

In the $\langle 110 \rangle$ symmetry direction (reached by a rotation by $45°$ along the surface normal) the first Brillouin zone crossing is two planes, (the (111) and $(11\bar{1})$ planes), making an angle of $\pm 54.35°$ with the $q_z = 0$ plane. Thus the band switches from 1 to 3 at the double crossing. This gives rise to the classic band gap behavior, band 1 having a minimum in binding energy (maximum in energy) at the Brillouin zone crossing, with its density petering out after the crossing. Band 3 slowly increases in intensity from zero momentum up to the first Brillouin zone boundary, where it has a maximum in binding energy, with increased density as one passes through Brillouin zone 3. The next extremum in energy, which corresponds to a minimum in binding energy, is when band 3 crosses the next set of Brillouin zone boundaries, i.e. on leaving Brillouin zone 3, and its intensity decreases thereafter as the momentum increases. The calculations and measurements are in quite good agreement with each other for these general features. For the intermediate angles band 1 remains dominant, band 4 is prominent, band 2 makes a significant contribution for directions not far from the $\langle 100 \rangle$ direction, and band 3 make small contributions close to the $\langle 110 \rangle$ direction.

Also shown in Figure 7.5 (bottom panels) is the spectral momentum density obtained along the $\langle 111 \rangle$ direction. This direction was reached by tilting the $\langle 110 \rangle$ aligned sample over $35.3°$ (see Figure 7.1(d)). Here the density is due to bands 1 and 2 with a large band gap at the zone crossing. Again these general features of the experiment and theory are in reasonable agreement. Note that the dispersion in the $\langle 111 \rangle$ direction could be mapped over a much larger momentum range than expected based on the momentum density distribution (see Figure 7.3 bottom panel). This is due to diffraction, as is described later.

A more detailed comparison between the calculated and measured dispersion of the bands in the three high symmetry directions is shown in Figure 7.6. Here a fitting procedure was used to extract the peak position of the measured density at different momenta. For comparison we include for the $\langle 100 \rangle$ and $\langle 111 \rangle$ directions also the ARPES data of Refs. [16] and [17]. The agreement between the calculation and the present measurement is excellent. There is a small deviation near the X-point in the $\langle 110 \rangle$ direction, with the observed gap between bands 1 and 3 being a little smaller than given by theory. A similar deviations for the band gaps is observed near the L points in the $\langle 111 \rangle$ direction. Generally the EMS shapes are smoother and better defined than the ARPES ones.

Figure 7.7 shows the band structures measured with the $\langle 110 \rangle$ direction aligned with the y-axis at three different values of q_x, i.e. 0, 0.65 a.u., and 0.87 a.u., the latter two being along Brillouin zone boundaries. The offsets in the x-component of momentum are produced by choosing suitable settings for the deflector voltages for each set of deflectors [22]. For the measurement with $q_x = q_z = 0$, there is a single dispersing feature with a band gap at $q_y = 0.65$ a.u. due to the crossing of the zone boundary. This momentum value corresponds to a K point of the band structure. Since the x- and y-directions through the Γ point are

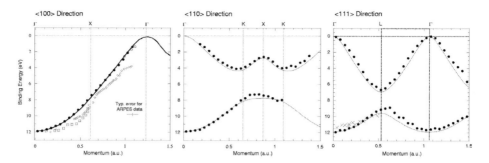

Figure 7.6: The measured dispersion in the peak density (dots) along the $\langle 100 \rangle$ (left panel), $\langle 110 \rangle$ (central) and $\langle 111 \rangle$ (right panel) symmetry directions. The full line is the LMTO calculation. Only those branches of the band structure are shown that are expected to have non-zero intensity in the EMS measurement. The open circles are ARPES data from ref. [16] and the diamonds, squares and crosses are ARPES data from [17].

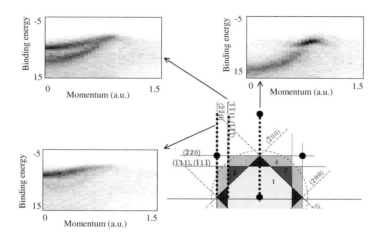

Figure 7.7: The measured densities along the lines indicated in the Brillouin zone cut. The x-axis of the spectrometer coincides with the crystal $\langle 1\bar{1}0 \rangle$ direction. One line goes through the Γ point (zero momentum), one has a constant q_x component of 0.65 a.u. and one has a constant q_x component of 0.87 a.u. The latter two lines coincide with Brillouin zone edges.

equivalent, the binding energy spectra at the two K points, $(q_x, q_y, q_z) = (0, 0.65, 0)$ and $(0.65, 0, 0)$, should be identical. This is indeed the case [22].

The total band width can be obtained from the zero momentum spectrum of the measurement along the $\langle 111 \rangle$ direction. Here both the top and the bottom of the band are clearly visible. The experimental separation between the two peaks is 11.85 ± 0.2 eV. The error is mainly due to the uncertainty in the shape of the background under the high-binding energy peak. The LMTO calculation gives a value of 11.93 eV.

As an alternative way of determining the valence band width we measured the valence band and Si 2p core level in the same experiment. This allows us to determine precisely the position of the valence band bottom relative to the $2p_{3/2}$ component of the core line. The fit result gives a value of 86.65 ± 0.2 eV for their energy separation. From photoemission data (e.g. Ref. [40]) we know that the $2p_{3/2}$ binding energy relative to the valence band maximum is 98.74 eV. In this way we obtain a value of 12.1 ± 0.2 eV for the valence band width.

7.4.2   Diffraction Effects

In a crystal coherent elastic scattering by the nuclear sites (dynamic diffraction) of the incident and/or the emitted electrons can shift the observed momentum distribution by a reciprocal lattice vector G [41–43]. In the present experiment the electron momenta are all very large ($k_0 = 62.1$ a.u., $k_{1,2} = 43.4$ a.u.), so that for the dominant diffraction the vectors G_i are all much smaller than the respective k_i. Thus the G_i must be essentially perpendicular to the k_i vector to fulfill the diffraction condition $2k_i \cdot G_i + G_i^2 = 0$. In the case where the surface normal (the $\langle 100 \rangle$ direction) is aligned with k_0 the smallest G_0 contributing to diffraction are of type $\langle \pm 2 \pm 2 \ 0 \rangle$ (see ref. [41] for a detailed discussion). The smallest vector q that can be accessed is 1.22 a.u. where the density is very low (see Figure 7.3), and it appears at $q_y = 1.22$ a.u. Diffraction events of the type $\langle 0 \pm 4 \ 0 \rangle$ produce an offset in the y-component of momentum by 2.44 a.u. The measured y-momentum line still passes through the origin so that we can still measure electrons with $q = 0$. Thus this diffracted beam causes a weak replication of the main spectrum shifted by 2.44 a.u. For the outgoing electrons the diffraction depends on which crystal direction is aligned with the y-axis of the spectrometer [41].

In the apparatus it is possible to rotate the sample about the y-axis. This does not affect the direction of measurement, but it does affect the directions of the incoming and outgoing electrons relative to the crystal symmetry directions and hence their diffraction. In general, rotating away from the symmetric configuration, there will be fewer small reciprocal lattice vectors that can contribute to diffraction, decreasing their influence. However, such a rotation cannot eliminate the contributions from reciprocal lattice vectors of the type $\langle 0 \pm 2 \ 0 \rangle$, which are perpendicular to the incoming and outgoing beams.

In Figure 7.8 we show the measured energy resolved momentum densities for two high symmetry directions in a grey-scale plot with the LMTO band structure in the repeated zone scheme superimposed on it. The contributions due to diffracted beams can be seen particularly at large momentum in the top panels with their enhanced grey-scale. At $q = 0$ the two spectral densities should be identical, however, as can be seen from the figure, at this Γ point there is a small peak in the density close to zero binding energy in the $\langle 100 \rangle$ direction, which is absent in the $\langle 110 \rangle$ direction. This density is due to the $\langle 111 \rangle$ reciprocal lattice vector (the shortest reciprocal lattice vector for the diamond lattice, length 1.06 a.u.) shifting density from $\Gamma_{(1,1,1)}$ to $\Gamma_{(1,0,0)}$. Rotation by $10°$ around the spectrometer y-axis removes this peak at low binding energies [41]. In this way, by examining the diffracted contributions in regions where the spectral density should be zero, it is possible to remove the diffracted components from the measured spectral momentum density. This is shown explicitly in Figure 7.9.

Diffraction effects are most obvious when the $\langle 111 \rangle$ direction is aligned with the spectrometer y-axis as can be seen already in the bottom panel of Figure 7.5. The measured intensity distribution is given by the crosses in Figure 7.9, which shows cuts through the data at the in-

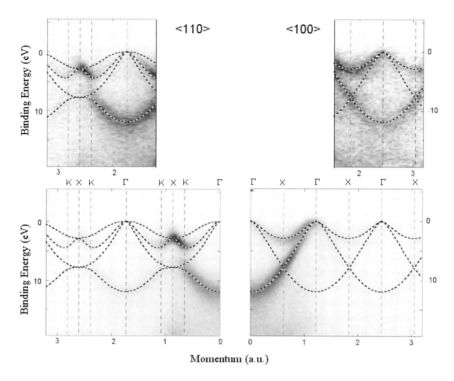

Figure 7.8: Measured momentum densities along the $\langle 110 \rangle$ (left) and $\langle 100 \rangle$ (right) directions in linear grey-scale with the LMTO band structure in the repeated zone scheme superimposed. The top panels show the high momentum parts with an enhanced grey-scale to highlight the low intensity diffracted contributions.

dicated binding energies. At the valence band maximum there are five peaks visible, at $q = 0$, ± 1.05 a.u., and ± 2.1 a.u. The dominant peaks at ± 1.05 a.u. are due to momentum density at the top of band 2 (see Figure 7.3), the other peaks are due to diffraction by a crystal reciprocal lattice vector. By assuming that the diffracted components add incoherently to the real spectral momentum density, the latter can be obtained by subtraction. The resulting momentum density is indicated by the solid line in Figure 7.9. The small gap between bands 1 and 2 (see Figure 7.3) shows up as a reduction in density around 8 eV.

After subtraction of the diffracted component the peak at the top of the valence band at $q = 0$ disappears completely, indicating that it was entirely due to diffraction with a reciprocal lattice vector of either $\boldsymbol{G} = \langle 111 \rangle$ or $\boldsymbol{G} = \langle -1 - 1 - 1 \rangle$. However, in the binding energy range near 6 eV there are clear shoulders left after subtraction. This is to be expected as near the Brillouin zone boundary these plane waves are expected to contribute significantly to the Bloch waves. The shoulder and main line that are part of the same Bloch function are shown in this figure, connected by a line of the length of the reciprocal lattice vector.

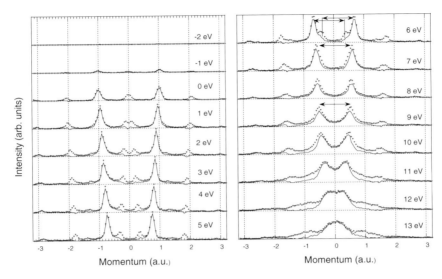

Figure 7.9: The measured intensities (crosses) along ⟨111⟩ direction. The solid lines give the momentum densities corrected for diffraction as discussed in the text. In the right panel the length of the reciprocal lattice vector $G = ⟨111⟩$ is indicated as well.

7.4.3 Many-body Effects

Up to this point we have concentrated on those aspects of the electronic structure of silicon that could be understood in the independent particle picture. However, even from the semi-quantitative grey-scale presentations in Figure 7.5 (and Figure 7.8) of the measured intensities it is clear that they do not follow the LMTO calculations in detail. The measurements show maximum intensity at intermediate energies or near the top of the band, whereas the theory predicts maximum intensity near the bottom of the band. Also the energy-widths of the measured density distributions at a given momentum are much larger than the theoretical ones even though the experimental energy resolution has been included in the calculations. This is due to lifetime broadening of the spectral momentum density by electron correlation effects. We will now look at some binding energy spectra at appropriate momenta in more detail in order to explore the role of many-body effects in the electronic structure of silicon.

Figure 7.10 shows two spectra obtained along the ⟨110⟩ direction, one at $q \simeq 0$ near the Γ point and the other at $q \simeq 0.87$ a.u. near the X point. The peak near the X point is very much narrower and taller than the peak near $q \simeq 0$. The latter is also quite asymmetric. As well as the raw data the figure includes a curve showing the data deconvoluted, in a parameter-free way, for inelastic scattering using only the measured energy loss spectrum (see Figure 7.2) [1]. This still leaves considerable intensity at high binding energies, extending some 20 eV or more above the position of the quasiparticle peak. If one tries to remove all the intensity at binding energies above the quasiparticle peak by further deconvolution one obtains non-physical negative intensities at some binding energies.

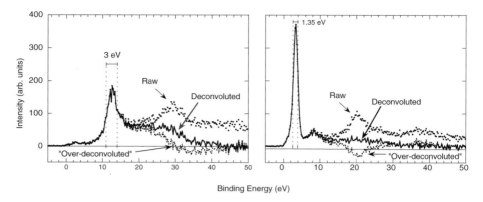

Figure 7.10: EMS spectra along the $\langle 110 \rangle$ direction at $q_y \simeq 0$ near the Γ point (left) and $q_y \simeq 0.87$ a.u. near the X point (right). An identical deconvolution procedure is followed for both spectra. The raw data are indicated by the dots, the data deconvoluted for inelastic multiple-scattering contributions by the solid line, and the squares show data that have deliberately been 'over-deconvoluted' (see text for details).

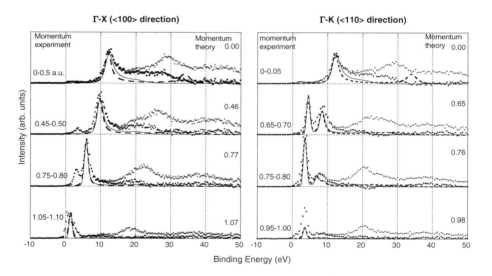

Figure 7.11: EMS spectra, raw and deconvoluted for inelastic multiple-scattering effects, at selected momenta along the $\langle 100 \rangle$ (top) and $\langle 110 \rangle$ (bottom) directions. The full and dashed lines are the results of the cumulant expansion and GW calculations, respectively, and they are normalized to the quasiparticle peak at $q = 0$.

In order to describe the data more quantitatively we performed full-scale many-body calculations using the GW and Cumulant Expansion approximations to the one-hole Green's function (see Section 7.3.2). The results are presented in Figure 7.11, which shows raw as

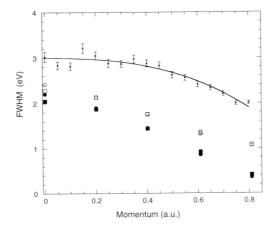

Figure 7.12: The FWHM of the quasiparticle peaks in the $\langle 100 \rangle$ direction plotted as a function of momentum. The measurements are indicated by the points with error bars. The thin line is a smooth fit to the experimentally obtained FWHM. The theoretical estimate of the peak width is indicated by squares (*GW* theory) and circles (cumulant expansion theory). The filled symbols refer to the calculated line width whereas the open symbols include the effect of finite experimental energy resolution.

well as the deconvoluted spectra along each of the two symmetry directions compared with the first-principles many-body calculations. The theories, convoluted with the experimental energy resolution, are normalized to the data at one point only, namely to the peak of the quasiparticle structure in the common $q \simeq 0$ spectra, the theories having the same total density in these spectra. The small binding energy peak present in the data in the $\langle 100 \rangle$ direction at intermediate q_y values and not in the calculations is due to the picking up of intensity from band 4 as discussed earlier.

The *GW* calculation gives a peak in the satellite intensity at around 34 eV at $q \simeq 0$. This is not observed in the data. This is a well-known failure of the GW approximation [13]. The cumulant expansion calculation gives a better fit to the data, although it too gives smaller satellite density than the experiment. Both calculations describe the main quasiparticle features quite well in all the spectra. In particular they give the broadening and the asymmetric structure of this feature reasonably accurately at the smaller momenta. It is this broadening which gives rise to the reduction in the peak heights at the lower momenta. However, even though the calculations give significant life-time broadening, they nevertheless underestimate the width of the quasiparticle structure, the cumulant expansion model giving a slightly better fit.

The finite quasiparticle (hole) lifetimes causes broadening of the observed features. In Figure 7.12 we plot as a function of the momentum the observed width (FWHM) for spectra obtained along the $\langle 100 \rangle$ direction. Near zero momentum the binding energy is a weak function of momentum, and the observed width should be a good indication of the life-time broadening. Away from zero momentum the binding energy becomes a strong function of momentum, and additional broadening is observed due to the finite momentum resolution of the spectrometer as initial states with slightly different momenta (and hence binding energy) contribute to the spectra. In spite of this a decrease in width is observed with increasing momentum. The sharpest spectra are observed where the dispersion goes again through an extremum (now minimum binding energy) i.e. near the X point ($\langle 100 \rangle$ direction, see also Figure 7.10) and the L point ($\langle 111 \rangle$ direction). Here a width of 1.35 eV is observed, a width clearly dominated by the energy resolution of the spectrometer. Also shown in the figure are the FWHM given by the many-body calculations. At zero momentum the calculated width is significantly smaller than the observed width (even if the calculations are broadened with the spectrometer energy resolution). The calculated width decreases more rapidly with momentum, which is, at least in part, due to the fact that we did not incorporate the effects of finite energy and momentum resolution in the calculations.

The density of the quasiparticle structure and the total density along the $\langle 100 \rangle$ direction are plotted in Figure 7.13. Also shown are the quasi-particle densities given by the cumulant expansion calculations, which somewhat underestimate the satellite density. For this figure the satellite density was defined as the total density above the main quasi-particle feature, whereas in the experiment we take the intensity extending 20 eV above the end of the quasi-particle peak. In this way we obtain for the experiment that the percentage of the total density that is contained in the satellite structure decreases as the momentum increases. For instance, along the $\langle 100 \rangle$ direction, the satellite structure has essentially disappeared by momentum values near 1 a.u., whereas at $q = 0$ it accounts for around 40% of the total density (Figure 7.13). In the theory the satellite intensity remains fairly constant (at 25% of the total density) but becomes more spread out over binding energy with increasing momentum.

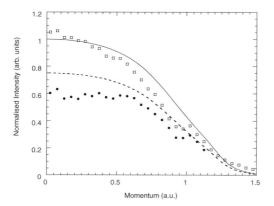

Figure 7.13: The energy integrated electron momentum density along the $\langle 100 \rangle$ direction. The open squares give the total density, and the solid circles the quasi-particle density. Also included is the total density given by the LMTO calculation (solid line) and the corresponding quasi-particle intensity, as predicted by the cumulant expansion calculations (dashed line).

7.5   Conclusions

We have measured the spectral momentum density of the prototype semiconductor Si along the three high symmetry directions, and several cuts between the $\langle 100 \rangle$ and $\langle 110 \rangle$ directions, all going through zero momentum (the $\Gamma_{(0,0,0)}$ point). In addition several measurements were made along the $\langle 110 \rangle$ direction but with the cuts shifted from $\Gamma_{(0,0,0)}$ point to coincide with nearby Brillouin zone edges.

The band structure, i.e. the dispersion of the position of the peak in the quasiparticle structure, is very well described by the FP-LMTO-DFT-LDA approximation. The present EMS data are in better agreement with the theory than the available ARPES data. The band-width was determined to be 11.85 ± 0.2 eV, in good agreement with ARPES data (12.5 ± 0.6 eV [44]) and the LMTO theory (11.93 eV).

Diffraction of the incident and/or the emitted electron beams can shift intensity from the 'direct' spectral density by reciprocal lattice vectors. Such effects are observed but can often be distinguished from the 'direct' intensity. Changing the diffraction conditions for the incoming and/or outgoing electrons pinpoints which part of the measured intensity is diffraction related. In this way contributions due to diffraction can be removed on the assumption that there is no interference between the direct and diffracted electron waves. The study of any possible interference effects could lead to additional information on the structure of silicon [41].

The spectra, especially at low momenta, show very large effects due to correlations. In particular the quasiparticle structures are very broad due to the short hole lifetimes, and much of the spectral density is found in higher energy satellite structure. This satellite structure is quite smooth and extends to around 20 eV above the quasiparticle peak. Although the satellite structure dominates the density at low momentum, its density drops off as the momentum is increased. Similarly the lifetime broadening diminishes as the momentum increases. Comparison with first principles many-body calculations shows that the *GW* approximation predicts the main quasiparticle features reasonably well, but cannot describe the satellite structure. The cumulant expansion approximation also describes the quasiparticle features qualitatively well, but does considerably better than the *GW* in its description of the shape of the satellite density. Although it predicts the energy distribution and the momentum dependence of the satellite structure quite well, it nevertheless significantly underestimates the satellite intensity, particularly at low momenta. The fact that agreement is better but still not perfect for the more detailed model is encouraging in that a quantitative comparison between the present EMS data and other representations of the many-body wavefunction should lead to new levels of understanding.

Acknowledgments

We are grateful to the Australian Research Council for financial support.

References

[1] M. Vos, A. S. Kheifets, V. A. Sashin, E. Weigold, M. Usuda, and F. Aryasetiawan, Quantitative measurement of the spectral function of aluminum and lithium by electron momentum spectroscopy, *Phys. Rev. B*, 66:155414, 2002.

[2] E. Weigold and I. E. McCarthy, *Electron Momentum Spectroscopy*, Kluwer Academic/Plenum, New York, 1999.

[3] M. Vos, G. P. Cornish, and E. Weigold, A high-energy (e,2e) spectrometer for the study of the spectral momentum density of materials, *Rev. Sci. Instrum.*, 71:3831–3840, 2000.

[4] M. Vos and E. Weigold, Developments in the measurement of spectral momentum densities with (e,2e) spectrometers, *J. Electron Spectrosc. Relat. Phenom.*, 112:93–106, 2000.

[5] A. Fleszar and W. Hanke, Spectral properties of quasiparticles in a semiconductor, *Phys. Rev. B*, 56:10228–10232, 1997.

[6] A.S. Kheifets and Y.Q. Cai, Energy-resolved electron momentum densities of diamond-structure semiconductors, *J. Phys.: Condens. Matter*, 7:1821–1833, 1995.

[7] W. Borrmann and P. Fulde, Exchange and correlation effects of the quasiparticle band structure of semiconductors, *Phys. Rev. B*, 35:9569–9579, 1987.

[8] K. Sturm, W. Schülke, and J. R. Schmitz, Plasmon-Fano resonance inside the particle-hole excitation spectrum of simple metals and semiconductors, *Phys. Rev. Lett*, 68:228–231, 1992.

[9] M. S. Hybertsen and S. G. Louie, Electron correlation in semiconductors and insulators: Band gaps and quasiparticle energies, *Phys. Rev. B*, 34:5390–5413, 1986.

[10] R. W. Godby, M. Schlüter, and L. J. Sham, Self-energy operators and exchange-correlation potentials in semiconductors, *Phys. Rev. B*, 37:10159–10175, 1988.

[11] M. Rohlfing, P. Krüger, and J. Pollmann, Quasiparticle band-structure calculations for C, Si, Ge, GaAs, and SiC using Gaussian-orbital basis sets, *Phys. Rev. B*, 48:17791–17805, 1993.

[12] G. E. Engel and W. E. Pickett, Investigation of density functionals to predict both ground-state properties and band structures, *Phys. Rev. B*, 54:8420–8429, 1996.

[13] L. Hedin, On correlation effects in electron spectroscopies and the GW approximation, *J. Phys: Condens. Matter*, 11:R489–R528, 1999.

[14] F. Aryasetiawan and O. Gunnarsson, The GW method, *Rep. Prog. Phys.*, 61:237, 1998.

[15] L. S. O. Johansson, P. E. S. Persson, U. O. Karlsson, and R. I. G. Uhrberg, Bulk electronic structure of silicon studied with angle-resolved photoemission from the Si(100) 2 x 1 surface, *Phys. Rev. B*, 42:8991–8999, 1990.

[16] A. L. Wachs, T. Miller, T. C. Hsieh, A. P. Shapiro, and T.-C. Chiang, Angle-resolved photoemission studies of Ge(111)-c(2x8), Ge(111)-(1x1)H, Si(111)-(7x7) and Si(100)-(2x1), *Phys. Rev. B*, 32:2326–2333, 1985.

[17] D. H. Rich, G. E. Franklin, F. M. Leibsle, T. Miller, and T. C. Chiang, Synchrotron photoemission studies of the Sb-passivated Si surfaces: Degenerate doping and bulk band dispersion, *Phys. Rev. B*, 40:11804–11816, 1989.

[18] A. Bansil, M. Lindroos, S. Sahrakorpi, R. S. Markiewicz, G. D. Gu, J. Avila, L. Roca, A. Tejeda, and M. C. Asensio, First principles simulations of energy and polarization dependent angle-resolved photoemission spectra of Bi2212, *J. Phys. Chem. Solids*, 63:2175–2180, 2002.

[19] F. Bell and J. R. Schneider, Three-dimensional electron momentum densities of solids, *J. Phys.: Condens. Matter*, 13:7905–7922, 2001.

[20] T. Sattler, Th. Tschentscher, J. R. Schneider, M. Vos, A. S. Kheifets, D. R. Lun, E. Weigold, G. Dollinger, H. Bross, and F. Bell, The anisotropy of the electron momentum density of graphite studied by $(\gamma,e\gamma)$ and (e,2e) spectroscopy, *Phys. Rev. B*, 63:155204, 2001.

[21] M. Itou, S. Kishimoto, H. Kawata, M. Ozaki, H. Sakurai, and F. Itoh, Three-dimensional electron momentum density of graphite by (x,ex) spectroscopy with a time of flight electron energy spectrometer, *J. Phys. Soc. Jpn.*, 68:515, 1999.

[22] M. Vos, V. A. Sashin, C. Bowles, A. S. Kheifets, and E. Weigold, Probing the spectral densities over the full three-dimensional momentum space, *J. Phys. Chem. Solids*, in press, 2004.

[23] S. J. Utteridge, V. A. Sashin, S. A. Canney, M. J. Ford, Z. Fang, D. R. Oliver, M. Vos, and E. Weigold, Preparation of a 10 nm thick single-crystal silicon membrane self-supporting over a diameter of 1 mm, *Appl. Surf. Sci.*, 162-163:359–367, 2000.

[24] M. Vos and M. Bottema, Monte Carlo simulations of (e,2e) experiments in solids, *Phys. Rev. B*, 54:5946–5954, 1996.

[25] M. Vos, A. S. Kheifets, and E. Weigold, Electron momentum spectroscopy of metals, In D. H. Madison and M. Schulz, editors, *Correlations, Polarization and Ionization in Atomic Systems, IAP Conference Proceedings 604*, pp. 70–75, American Institute of Physics, New York, 2002.

[26] R.O. Jones and O. Gunnarsson, The density functional formalism, its applications and prospects, *Rev. Mod. Phys.*, 61:689–746, 1989.

[27] R.M. Dreizler and E.K.U. Gross, *Density Functional Theory*, Springer Verlag, Berlin, 1990.

[28] W.E. Pickett, Electronic structure of the high-temperature oxide superconductors, *Rev. Mod. Phys.*, 61:433–512, 1989.

[29] H.L. Skriver, *The LMTO Method*, Springer Verlag, Berlin, 1984.

[30] A. S. Kheifets, D. R. Lun, and S. Yu Savrasov, Full-potential linear-muffin-tin-orbital calculation of electron momentum densities, *J. Phys.: Condens. Matter*, 11:6779–6792, 1999.

[31] L. Hedin, Method for calculating the one-particle green's function with application to the electron-gas problem, *Phys. Rev.*, 139:A796–A823, 1965.

[32] L. Hedin and S. Lundqvist, *Solid State Phys.*, 23:1–181, 1969.

[33] D. C. Langreth, Singularities in x-ray spectra of metals, *Phys. Rev. B*, 1:471, 1970.

[34] B. Bergersen, F. W. Klus, and C. Blomberg, Single particle Green's function in the electron-plasmon approximation, *Can.J.Phys.*, 51:102, 1973.

[35] L. Hedin, Effects of recoil on shake-up spectra in metals, *Phys. Scr.*, 21:477–480, 1980.

[36] F. Aryasetiawan, L. Hedin, and K. Karlsson, Multiple plasmon satellites in Na and Al spectral functions from ab initio cumulant expansion, *Phys. Rev. Lett.*, 77:2268–2271, 1996.

[37] M. Vos, A. S. Kheifets, E. Weigold, S. A. Canney, B. Holm, F. Aryasetiawan, and K. Karlsson, Determination of the energy-momentum densities of aluminium by electron momentum spectroscopy, *J. Phys.: Condens. Matter*, 11:3645, 1999.

[38] M. Vos, A. S. Kheifets, and E. Weigold, The spectral momentum density of aluminum measured by electron momentum spectroscopy, *J. Phys. Chem. Solids*, 62:2215–2221, 2001.

[39] V. Heine, *Group Theory in Quantum Mechanics*, Pergamon, New York, 1960.

[40] F.J. Himpsel, G. Hollinger, and R.A. Pollak, Determination of the Fermi-level pinning position at Si(111) surfaces, *Phys. Rev. B*, 28:7014–7018, 1983.

[41] M. Vos, A. S. Kheifets, V. A. Sashin, and E. Weigold, Influence of electron diffraction on measured energy-resolved momentum densities in single-crystalline silicon, *J. Phys Chem. Solids*, 64:2507–2515, 2003.

[42] L. J. Allen, I. E. McCarthy, V. W. Maslen, and C. J. Rossouw, Effects of diffraction on the (e,2e) reaction in crystals, *Aust. J. Phys.*, 43:453–464, 1990.

[43] R.S. Matthews, PhD Thesis, Flinders University of South Australia, 1993.

[44] W.D. Grobman and D.E. Eastman, Photoemission valence-band densities of states for Si, Ge, and GaAs using synchrotron radiation, *Phys. Rev. Lett.*, 29:1508–1512, 1972.

8 Recent Results from $(\gamma, e\gamma)$ and Compton Spectroscopy

Friedhelm Bell

We report on the measurement of the three-dimensional electron momentum density (EMD) of graphite, fullerenes and a Cu–Ni alloy by the $(\gamma, e\gamma)$ reaction. Experiments have been performed either at the wiggler beamline of the European Synchrotron Radiation Facility (ESRF) or the undulator beamline of the PETRA storage ring at Deutsches Elektronen-Synchrotron (DESY) with photon energies of about 150 keV. The influence of lifetime effects due to final state interactions is discussed.

8.1 Introduction

The strong interest in the electronic structure of solids led to the development of a large variety of experimental methods for the study of energy dispersion and density of states both for occupied and unoccupied bands: photoemission spectroscopy [1], inelastic X-ray scattering [2], electron-loss spectroscopy [3], X-ray absorption spectroscopy [4], to name but a few. In contrast, a few methods exist which measure directly wave function-related quantities such as the real space electron density (X-ray form factors) or momentum densities. To the latter belong the two-dimensional angular correlation of annihilation radiation (2D-ACAR) [5] — which, strictly speaking, measures the electron–positron pair density $\rho^{2\gamma}(\boldsymbol{p})$ [5], — $(\gamma, e\gamma)$ [6] and (e,2e) [7] spectroscopy. $(\gamma, e\gamma)$ experiments which are reviewed in this article are an extension of conventional Compton scattering where the double differential cross section describing the energy and angular distribution of the scattered radiation is proportional to the so–called Compton profile (CP) $J(p_z)$, which is defined as a twofold integration over the electron momentum distribution (EMD) $\rho(\boldsymbol{p})$:

$$J(p_z) = \iint \rho(\boldsymbol{p})\, \mathrm{d}p_x \mathrm{d}p_y \tag{8.1}$$

This integration results from the lack of information about the momentum distribution of the recoiling electrons. It is therefore desirable to measure the EMD directly by fixing the complete scattering kinematics: if the momenta of the primary and scattered photon in addition to that of the recoil electron are measured simultaneously, i.e. in coincidence, the momentum of the electron in its initial state can be determined in a unique way. The corresponding triple differential cross section is proportional to the EMD itself. The main difficulty of such an experiment originates in the strong incoherent elastic scattering of the recoiling electron within the target which disturbs the determination of the recoil momentum by multiple scattering.

Correlation Spectroscopy of Surfaces, Thin Films, and Nanostructures. Edited by Jamal Berakdar, Jürgen Kirschner
Copyright © 2004 Wiley-VCH Verlag GmbH & Co. KGaA, Weinheim
ISBN: 3-527-40477-5

A method very similar to $(\gamma, e\gamma)$ is the (e,2e) reaction where instead of a photon an energetic electron is used as projectile [7]. Again, multiple electron scattering within the target is the most severe problem. Comparing $(\gamma, e\gamma)$ and (e,2e) experiments the situation is more relaxed in the former case since at least the photon will not be multiply scattered. The advantage of the (e,2e) technique is its large cross section (Mott compared to Klein–Nishina) and its monochromatic projectile flux. A highly monochromatic electron flux of 10^{12} electrons s^{-1} (100 nA) is easily achieved whereas comparable photon fluxes with considerably less monochromaticity are obtained only from synchrotron radiation facilities of the third generation. Altogether, this allows the introduction of electron spectrometers in (e,2e) experiments which in turn makes it possible to measure the EMD of solids as a function of the valence binding energy with a resolution of about 1.0 eV [7], which enables one to measure not only the EMD $\rho(\boldsymbol{p})$ but also the spectral function $A(\boldsymbol{p}, E)$ from which the EMD can be derived

$$\rho(\boldsymbol{p}) = \sum_{\boldsymbol{k},\boldsymbol{g},j} n_j(\boldsymbol{k}) |C_j(\boldsymbol{p})|^2 \delta_{\boldsymbol{p},\boldsymbol{k}+\boldsymbol{g}} \tag{8.2}$$

where the $C_j(\boldsymbol{p})$ are the plane wave coefficients in the development of the Bloch wave $|j\boldsymbol{k}\rangle$

$$|j\boldsymbol{k}\rangle = \sum_{\boldsymbol{g}} C_j(\boldsymbol{k}+\boldsymbol{g}) |\boldsymbol{k}+\boldsymbol{g}\rangle \tag{8.3}$$

and the occupation number density $n_j(\boldsymbol{k})$ is given by the spectral function [8]

$$n_j(\boldsymbol{k}) = \int_{-\infty}^{E_{\mathrm{F}}} A_j(\boldsymbol{k}, E) \frac{\mathrm{d}E}{2\pi} \qquad\qquad \boldsymbol{k} \in 1.BZ \tag{8.4}$$

$\boldsymbol{g}$ is a reciprocal lattice vector and E_{F} the Fermi energy. Usually, energy resolution in the case of $(\gamma, e\gamma)$ is not good enough to be selective to specific initial states j, i.e. the EMD is summed over all j. In the following we will use atomic units, i.e. $\hbar = m = e = 1$.

8.2   Experiment

The $(\gamma, e\gamma)$ experiments were performed either at the high-energy X-ray undulator beamline of HASYLAB at the 12 GeV PETRA storage ring or at the wiggler beamline ID15A of the 6 GeV storage ring of the ESRF. Since the experimental set-ups are not very different we describe in the following that of the PETRA experiment only. The white photon beam was monochromatized by a plane, slightly disordered Si crystal in Laue geometry [9]. The disorder widens the rocking curve considerably compared to the Darwin width of a perfect crystal and matched the monochromaticity of the photon beam (width $\sigma_\omega = 0.35$ keV) with the energy resolution of the photon detector (see below) without a major loss of reflectivity. Thus, a photon flux of 2×10^{12} photons s^{-1} in a beam spot of $2 \times 2\,\mathrm{mm}^2$ could be reached at a photon energy of 180.3 keV and an average storage ring current of 30 mA. For photon detection we implemented a two-dimensional array of 12 intrinsic Ge diodes (energy resolution $\sigma_{\omega'} = 0.32$ keV) which was mounted externally to the evacuated target chamber at a scattering angle of $\theta = 150°$. The electrons were identified by a position-sensitive detector (PSD) consisting of a two-dimensional array of 32 individual PIN diodes. (Figure 8.1).

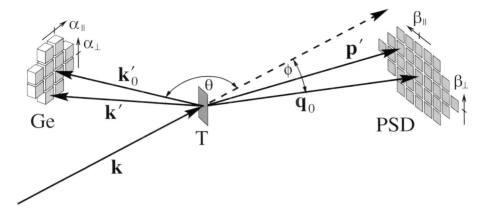

Figure 8.1: Experimental $(\gamma, e\gamma)$ setup: Ge, 12-pixel Ge diode; T, target; PSD, 32-pixel position sensitive electron detector.

8.3   Results and Discussion

8.3.1   Graphite

We have used two modifications of graphite, made either by thermal evaporation (te) or laser plasma ablation (lpa). For detailed information about both the production procedure and later heat treatment of the foils we refer to Ref. [10]. Both kinds of foils had a thickness of about 18 nm. For the te-foils the basal planes of graphite were arranged mostly parallel to the foil surface but were shifted and rotated randomly to each other. In contrast the lpa-foils showed an isotropic distribution of crystallites.

In the following experimental EMDs will be compared with theoretical ones obtained from band structure calculations [10]. For that the experimental data have been normalized to an effective number of electrons by integrating both experiment and theory over the same volume in momentum space which is determined by experimental conditions. Theoretical EMDs are based on either an empirical pseudopotential (PP) method [11] with potential parameters from Reed et al. [12], the full-potential linear muffin-tin orbital (FP-LMTO) [13] or the modified augmented plane wave (MAPW) method [14]. All calculations were performed within the general scheme of density functional theory (DFT). Due to the structure of the foils observed by electron diffraction the theoretical EMDs have been either spherically averaged to represent the lpa-targets or azimuthally averaged in the case of the te-foil. The momentum component parallel to the c-axis (surface normal of the basal plane) is called $p_{||} = p_z$, the component within the basel plane $p_\perp = \sqrt{p_x^2 + p_y^2}$.

A qualitative demonstration of the anisotropy of the graphite EMD is shown in Figure 8.2(a) where contour lines of the difference of the theoretical EMDs, i.e., basal averaged minus spherically averaged EMD, are plotted as a function of $p_\perp$ and $p_{||}$. Figure 8.2(b) shows the influence of resolution and electron multiple scattering on this difference. Evidently, the strength of the anisotropies is reduced, but the general structure is retained. In Figure 8.2(c)

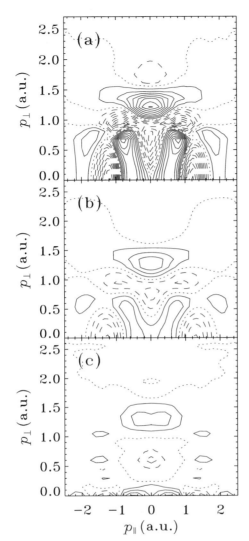

Figure 8.2: The difference of azimuthally minus spherically averaged EMD. Solid lines hold for a positive difference, dashed lines for a negative one, and the dotted line refers to zero. The theoretical FP-LMTO difference (a), the theoretical difference corrected for experimental resolution and electron multiple scattering (b), and the experimental difference (c). The latter has been symmetrized with respect to $p_{\parallel} = 0$. The difference between the lines is 0.005 a.u.$^{-3}$ in each case.

the experimental difference is plotted and should be compared with the theoretical difference of Figure 8.2(b). The comparison reveals that the general features of the experimental anisotropies are reproduced by theory. In a more quantitative manner Figure 8.3 shows the difference $\Delta\rho$ of the EMDs from te- and lpa-foils as a function of $p_{\parallel}$ at certain $p_{\perp}$-values (Figure 8.3, left column) or as a function of $p_{\perp}$ for constant $p_{\parallel}$ (right column). Experimental data are compared again with PP (broken curve), FP-LMTO (solid curve) and MAPW (dash-dotted) calculations. Again, the anisotropy is better reproduced by the FP-LMTO and MAPW calculations rather than the PP one.

We mention that PP calculation of Lou [11] also failed to describe the non-coincident Compton scattering experiment on pyrolytic graphite of Manninen et al. [15]. The sequence of

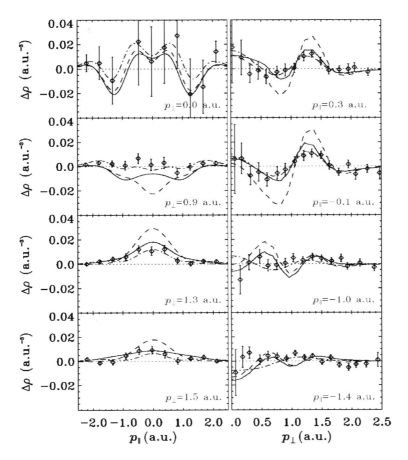

Figure 8.3: The difference $\Delta\rho$: the EMD of the te-foil minus that of the lpa-foil. PP: broken line; FP-LMTO: solid line; MAPW: dash-dotted line. $\Delta\rho$ as a function of $p_{\parallel}$: left column; $\Delta\rho$ as a function of $p_{\perp}$: right column

theories if compared with experiment and shown by the 1D cuts of Figure 8.3 is also supported by the global fits of the complete 3D-EMD: the normalized χ^2/n (n: number of bins) is 2.15 (MAPW), 3.07 (FP-LMTO) and 3.80 (PP) in the case of the te-foil, and 2.22 (MAPW), 2.40 (FP-LMTO) and 2.71 (PP) for the lpa-foil.

8.3.2 Fullerene

The most remarkable difference in the EMDs of graphite and C_{60} is the increased density of C_{60} compared to graphite at small momenta [16]. The bending of the graphite basal planes to form the buckyball induces a hybridization of wavefunctions with s-character with those of the π-electrons, which results in a charge transfer from the interlayer region into the shell of the C_{60} spheres [17]. (Nevertheless, bonding is closer to the sp^2 hybrids of graphite than to the sp^3

bonding in diamond [18]). Consequently, a DFT calculation in local density approximation (LDA) revealed that the electron density for most of the 120 bands in fullerene showed a larger overlap among the nearest neighbour carbon atoms in the C_{60} ball than in graphite [19]. This kind of delocalization results apparently in an enhancement of the EMD at small momenta. A comparison of one-dimensional cuts through the EMD of fullerene with either the FP-LMTO for graphite or the molecular C_{60} calculation showed that experiment could not distinguish between both theories within error bars [16]. To improve the statistics we have summed all coincidence events for a constant p_z-value. The resulting coincident Compton profile $J_{coinc}(p_z)$ is not identical with a non-coincident one due to the limited integration range in the p_x and p_y directions which results from the finite extension of the electron detector, but besides the increase in statistics it also has the advantage that measurements in coincidence provide photon spectra free from any background radiation. Again, experimental J_{coinc} have been normalized to the effective number of electrons given by the restricted range of momenta.

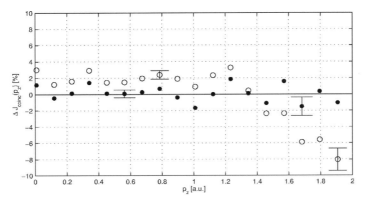

Figure 8.4: The relative difference ΔJ_{coinc} for the FP-LMTO calculation for graphite (open circles) and the C_{60} calculation (dots).

To enlarge possible differences we have plotted in Figure 8.4 the relative difference $\Delta J_{coinc} = (J_{exp} - J_{theor})/J_{exp} \times 100$ where J_{exp} is the experimental coincident Compton profile and J_{theor} the corresponding theoretical profile, i.e. the theoretical EMD has been integrated over the limited (p_x, p_y) range mentioned above. Open circles are from FP-LMTO for graphite, full circles represent the molecular C_{60} calculation. Though the relative difference is rather small experiment clearly favours the C_{60} calculation. Figure 8.4 also demonstrates the above mentioned momentum transfer towards smaller momenta if C_{60} is formed from graphene sheets: due to norm conservation this is reflected by $J_{exp}(C_{60}) < J_{theor}(\text{graphite})$ for large momenta.

8.3.3   Cu–Ni Alloy

In this chapter we will report on the influence of alloying by comparing the EMD of $Cu_{0.50}Ni_{0.50}$ with those of its pure constituents [20]. The electronic structure of this 3d transition-metal solid-solution alloy has received extensive attention. The Cu–Ni system is

completely soluble over the whole concentration range, i.e. there is no miscibility gap yielding single phase disordered fcc alloys. Substantial improvement has been achieved in the
theoretical understanding of the electronic structure of disordered alloys using a general multiple scattering formalism. The underlying Green's function technique is similar to that of
the Korringa–Kohn–Rostoker theory (KKR). In the coherent potential approximation (CPA)
self-consistency is required to obtain the single site t-matrix [21]. It is this property of the
CPA which makes it preferable to apply, especially for alloys of high concentrations.

Since self-supporting Cu or Ni foils with diameters of 8 mm and thicknesses of 20 nm
cannot be prepared we have evaporated 22 nm Cu followed by 22 nm Ni on a 30 nm thin C-
foil acting as a backing. The backing was made by condensation of evaporated carbon atoms
on a thin betaine film which finally was dissolved in water, and the carbon film was mounted
on a frame. Both Cu and Ni were evaporated on the free-standing carbon backing at a rate of
about 0.5 nm s^{-1}, and the films condensed at room temperature. We first investigated the
sandwich foil and then heated it to about 500 °C for 2 h. That in fact an alloy has been formed
by this heat treatment was confirmed by X-ray diffraction experiments made either with the
sandwich or the annealed specimen [6]. Finally we measured the EMD of the alloy.

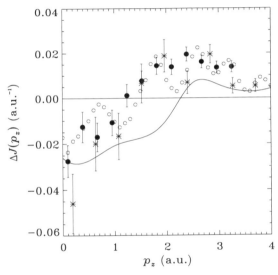

Figure 8.5: The coincident Compton profile difference ΔJ (stars from ESRF, dots from DESY)
and the non-coincident CP of [22] (open circles). The solid curve represents KKR-CPA theory [22].

To improve the statistics we have again summed all events for a constant p_z to obtain
a coincident Compton profile J_{coinc}, which has been evaluated in absolute units by normal-
izing it to the effective number Z_{eff} of electrons contributing to it. Figure 8.5 shows the
difference $\Delta J = J_{\text{coinc}}(\text{sandwich}) - J_{\text{coinc}}(\text{alloy})$ which demonstrates deviation from simple additivity of EMDs. Due to the improved statistics, an alloying effect is clearly observ-
able. At $p_z = 0$ it amounts to about 2 % of the total coincident profile, i.e. it is a rather
small effect. We also compare in Figure 8.5 the experimental non-coincident CP-difference

$\Delta J = (J_{\mathrm{Cu}} + J_{\mathrm{Ni}})/2 - J_{\mathrm{Cu}_{0.5}\mathrm{Ni}_{0.5}}$ [22]. Despite the oscillations the general trend of the experimental points (open circles) agrees nicely with our results. The oscillatory behavior might be an artifact introduced by the numerical procedure of data processing [22]. Neither data processing like a deconvolution procedure accompanied by frequency filtering nor background subtraction has been applied to our data. The solid line in Figure 8.5 represents KKR-CPA theory [20, 22]. In view of the smallness of the effect a reasonable agreement between theory and experiment is observed. We emphasize that both experiment and theory are on an absolute scale in Figure 8.5.

8.4   Lifetime Effects in Compton Scattering

Recent conventional Compton scattering investigations are so-called high-resolution experiments where both the initial and scattered X-ray beam are analyzed by a Bragg-spectrometer. Thus, p_z-resolutions of about 0.02 a.u. can be achieved [23, 24]. A major issue in recent years has become the investigation of nearly free electron (NFE)-metals like Li or Be. It was one of the milestones in the investigation of many-body effects for the electron gas, that the Fermi break remains sharp if electron–electron interaction is allowed for (Luttinger's theorem [25]), an effect which results from the vanishing of the inverse lifetime at the Fermi momentum p_{F}. Whereas the break remains sharp the height of the Fermi-edge is reduced from one to the renormalization constant Z_{F}. At the same time EMD intensity appears above p_{F} which results from satellites due to plasmaron excitation [26].

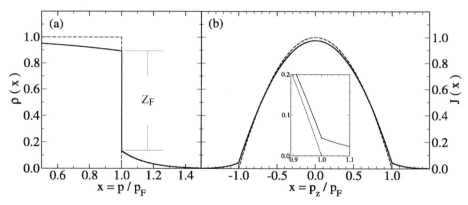

Figure 8.6: (a) The EMD $\rho(x)$ and (b) the Compton profile J(x) for both the non-interacting (broken line) and the interacting electron gas (solid line). The insert shows the enlarged CPs at the Fermi momentum.

Figure 8.6(a) shows the EMD for both a non-interacting and an interacting electron gas together with the corresponding CPs, Figure 8.6(b). Strictly speaking, $\rho(x)$ is the occupation number density of Eq. (8.4) since for the electron gas the wavefunctions are plane waves, i.e. $|C_j(\mathbf{p})|^2 = \delta_{g,0}$ in the empty lattice approximation. This means that the simple picture of an occupancy for every state below the Fermi energy equal to one becomes invalid. Since most of the band structure calculations use the occupancy of the non-interacting electron gas one has to correct EMDs, a procedure known as the Lam–Platzman correction [27]. Due to the sharp break in the EMD the CP has a discontinuity at p_F, see the insert of Figure 8.6(b). Of course, in any experiment this discontinuity will be smeared out due to finite experimental resolution. In addition even a weak electron–ion interaction in a NFE-metal distorts the EMD of Figure 8.6(a). Nevertheless, experimental results indicated a fundamental problem: Figure 8.7 shows the <111> CP for Li and the insert an enlarged view at the Fermi-momentum: the theoretical result from a KKR-LDA band structure calculation (dotted curve) is far from being a good description of the experimental results (dots with error bars [28]). These findings seemed to indicate a remarkable stronger smearing of the Fermi break than assumed until now or, with respect to Figure 8.6(a), could also be interpreted by a renormalization constant Z_F close to zero [29]. But about 40 years of intensive investigations of many-body theory – one should not forget that the interacting electron gas is a benchmark in many-body solid state theory – demanded that Z_F is about 0.7 ± 0.15 in Li [30].

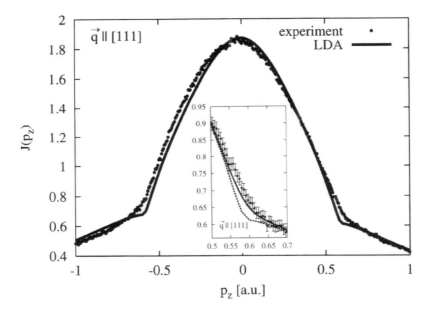

Figure 8.7: The <111> CP of Li, experiment (dots) and theory (solid line). The insert shows an enlarged view at the Fermi momentum: experiment (dots with error bars [28]); KKR-LDA theory (dotted curve); KKR-LDA folded with final state interactions (solid curve)

The solution of this fundamental problem proposed by Sternemann et al. [23] was to consider not only lifetime broadening in the initial state – which had been considered up to now – but also in the final state due to inelastic scattering of the recoiling electron. It is thus assumed that for the measured CP [2, 31]

$$J_{\mathrm{meas}}(\boldsymbol{q}, \omega) \propto S(\boldsymbol{q}, \omega) = \iint \Xi(\boldsymbol{p}, E) \frac{\mathrm{d}^3 \boldsymbol{p}}{(2\pi)^3} \frac{\mathrm{d}E}{2\pi} \tag{8.5}$$

holds with the joint spectral function (JSF)

$$\Xi(\boldsymbol{p}, E) = A_i(\boldsymbol{p}, E) A_f(\boldsymbol{p} + \boldsymbol{q}, E + \omega) \tag{8.6}$$

and $\boldsymbol{q}$ and ω as the momentum and energy transfer respectively. $S(\boldsymbol{q}, \omega)$ is the so-called dynamical structure factor. If the quasiparticle character of the recoiling electron could be neglected one has for an infinitely long lifetime and with the bare particle energy $\epsilon_{\boldsymbol{p}}$

$$A_f = 2\pi\delta(\epsilon_{\boldsymbol{p}+\boldsymbol{q}} - \epsilon_{\boldsymbol{p}} - \omega) = 2\pi\delta(q^2/2 + p_z q - \omega) \tag{8.7}$$

which yields according to Eq. (8.2) $J_{\text{meas}} = J(p_z)$. But allowing for finite lifetimes and assuming, for simplicity, a Lorentzian spectral function one obtains

$$J_{\text{meas}}(p_z) = \int\limits_{-\infty}^{+\infty} J(p_z') \frac{1}{\pi} \frac{\Delta p}{(p_z' - p_z)^2 + (\Delta p)^2} \, dp_z' \tag{8.8}$$

i.e. a Lorentzian smearing of the CP due to final state interactions. Lifetime effects in the initial state are thought to be incorporated in $J(p_z)$. The smearing Δp is determined by the inverse final state lifetime Γ_f [32]

$$\Delta p = \frac{\Gamma_f}{q} = \frac{1}{2\lambda(q)} \tag{8.9}$$

where $\lambda(q)$ is the mean free path for inelastic scattering of an electron with momentum q. For a momentum transfer $q = 5$ a.u. [23] one has $\lambda \simeq 6.3$ a.u. or $\Delta p \simeq 0.08$ a.u. [32] which is considerably larger than the experimental resolution of 0.02 a.u. If the theoretical KKR-LDA Compton profile is folded with the Lorentzian of Eq. (8.8) one obtains the solid curve in the insert of Figure 8.7 which improves the agreement with the experimental date considerably.

In the following we extend such an analysis to $(\gamma, e\gamma)$ experiments, assuming for the moment a sufficiently good energy and momentum resolution. In this sense the analysis may also apply to (e,2e) experiments. Neglecting final state interactions the triply differential cross section of a $(\gamma, e\gamma)$ interaction is proportional to the spectral function $A(\boldsymbol{p}, E)$. But for a finite lifetime of the recoil electron we assume that according to the reasoning given above the cross section will depend on JSF. Actually, the spectrum measured will be a 1D-cut through the four-dimensional $(\boldsymbol{p}, E)$ manifold. This will either be an energy scan (ES) for fixed momenta, or a momentum scan (MS) with constant energy. Considering ES first we obtain for the resulting intensity distribution I_{ES}

$$I_{\text{ES}}(E) \propto \int \Xi(p, E) \, dp \tag{8.10}$$

Assuming again Lorentzians for both A_i and A_f and developing the band dispersion up to first order [33] one obtains

$$I_{\text{ES}}(E) \propto \frac{1}{(E - \epsilon(p_0))^2 + (\Delta E)^2} \tag{8.11}$$

with an energy lifetime broadening (ELB)

$$\Delta E = \frac{v_{\text{f}} \Gamma_{\text{i}} + v_{\text{i}} \Gamma_{\text{f}}}{|v_{\text{f}} - v_{\text{i}}|} \tag{8.12}$$

The Γ_μ are the inverse lifetimes of the initial and final state (hole and electron states, resp.) and the v_μ corresponding group velocities [33]. Equation (8.12) is well known from photo-emission spectroscopy [34].

Alternatively, one gets for MS

$$I_{MS}(p) \propto \int \Xi(p, E)\, dE \tag{8.13}$$

These scans hold for the so-called Doppler-broadening mode [10]. In this case one obtains

$$I_{MS}(p_z) \propto \frac{1}{(p_z - p_z')^2 + (\Delta p)^2} \tag{8.14}$$

where p_z is the momentum component parallel to q and $p_z' = (q^2/2 - \omega)/q$. The momentum lifetime broadening (MLB) reads $\Delta p = (\Gamma_i + \Gamma_f)/q$.

In the following we shortly discuss consequences of such an analysis:

1. ELB: for valence states at the Fermi level ($\Gamma_i = 0$) one has approximately $\Delta E = (v_F/v_f)\Gamma_f$. Since for both $(\gamma, e\gamma)$ and (e,2e) experiments $v_F/v_f \ll 1$ holds ELB will be negligible [7]. For flat dispersion curves, as in the case of core states, valence states in insulators or at the bottom of the valence band one has $v_i \approx 0$, i.e. ELB is solely determined by the initial state: $\Delta E \simeq \Gamma_i$ [7]

2. MLB: for valence states at the Fermi-level one has $\Delta p = \Gamma_f/q$. This is the result derived earlier and explains the Compton scattering experiments for Li and Be [32].

It might also be possible to observe lifetime broadening due to final state interactions in electron energy-loss spectroscopy (EELS). Of course, this would only hold for single-particle excitations and mixed modes as the plasmaron (intrinsic plasmon satellites [7]) but not for pure collective excitations. In the case of core states the same reasoning as mentioned above yields that final lifetime broadening can be neglected. Interband or intraband transitions for valence states are very weak, at least if compared with neighboring plasmon peaks and are placed on a strong background. A line shape analysis in terms of initial state lifetime broadening (plasmon damping) yielded partial agreement [3, 35], but the influence of final state interactions on the line shape cannot be excluded.

Finally, we remark that we have always assumed that $\Delta E \ll E$, since otherwise lifetime broadening effects would be influenced by energy conservation laws. It is well known from resonant inelastic X-ray scattering (RIXS) [36], angle-resolved photoemission spetroscopy (ARPES) [37] or Auger photoelectron coincidence spectroscopy (APECS) [38] that substantial lifetime narrowing may occur in these cases.

8.5 Summary

We have shown that by coincident detection of the scattered X-ray photon and its recoil electron the complete 3D electron momentum density of solids can be measured. Examples for graphite and fullerene are given and the experimental EMDs are compared with theoretical ones obtained from band structure calculations. In the case of graphite FP-LMTO and MAPW results describe the experimental EMD better than a pseudo-potential calculation.

For fullerenes the bending of graphene sheets to form a buckyball changes the EMD in a characteristic way which is revealed by experiment. The influence of alloying on the EMD was studied in the Cu–Ni system and the results are in satisfactory agreement with a KKR-CPA theory. Lifetime effects due to final state interactions are discussed.

Though the results presented above are extremely important we have little hope of getting the Nobel Prize for it since that was given already for another $(\gamma, e\gamma)$-experiment published by Bothe and Geiger in 1925 [39].

Acknowledgment

We are indebted to our colleagues who contributed over the years to these investigations: F. Kurp, C. Metz, A. J. Rollason, T. Sattler, J. R. Schneider and Th. Tschentscher. We wish to thank H. Bross, A. S. Kheifets and Y. Lou who provided us with their band structure calculations. This work was supported by the German Federal Ministry of Education and Research, Contract Nos 055WMAAI and 05ST8HRA.

References

[1] H. Levinson, F. Greuter, and E. Plummer, Phys. Rev. B **27**, 727 (1983).

[2] W. Schülke, J. Phys.: Condens. Matter **13**, 7557 (2001).

[3] J. Fink, M. Knupfer, S. Atzkern,and M. S. Golden J. Electron Spectrosc. Relat. Phenom. **117-118**, 287 (2001).

[4] I. Batra and L. Kleinman, J. Electron Spectrosc. Relat. Phenom. **33**, 175 (1984).

[5] M. Puska and R. Nieminen, Rev. Mod. Phys. **66**, 841 (1994).

[6] F. Bell and J. R. Schneider, J. Phys.: Condens. Matter **13**, 7905 (2001).

[7] M. Vos, A.S. Kheifets, V.A. Sashin, E. Weigold, M. Usuda, and F. Aryasetiawan Phys. Rev. B **66**, 155414 (2001).

[8] A.G. Eguiluz, W. Ku, and J.M. Sullivan, J. Phys. Chem. Solids **61**, 383 (2000).

[9] J.R. Schneider, O.D. Gonçalves, A.J. Rollason, U. Bonse, J. Lauer, and W. Zulehner, Nucl. Instrum. Methods B **29**, 661 (1988).

[10] T. Sattler, Th. Tschentscher, J. R. Schneider, M. Vos, A. S. Kheifets, D. R. Lun, E. Weigold, G. Dollinger, H. Bross, and F. Bell, Phys. Rev. B **63** 155204 (2001).

[11] Y. Lou, B. Johansson, and R.M. Nieminen, J. Phys.: Condens. Matter **3**, 1699 (1991).

[12] W.A. Reed, P. Eisenberger, K.C. Pandey, and L.C. Snyder, Phys. Rev. B **10**, 1507 (1974).

[13] A.S. Kheifets, D.R. Lun, and S. Yu Savrasov, J. Phys.: Condens. Matter **11**, 6779 (1999).

[14] H. Bross, G. Bohn, G. Meister, W. Schubö, and H. Stöhr, Phys. Rev. B **2**, 3098 (1970).

[15] S. Manninen, V. Honkimäki, and P. Suortti, Z. Naturforsch. **48a**, 295 (1993).

[16] C. Metz, Th. Tschentscher, P. Suortti, A.S. Kheifets, D.R. Lun, T. Sattler, J.R. Schneider, and F. Bell, J. Phys.: Condens. Matter **11**, 3933 (1999).

[17] N. Troullier and J. L. Martins, Phys. Rev. B **46**, 1754 (1992).

[18] R. Taylor and D. R. M. Walton, Nature **363**, 685 (1993).

[19] J. Moscovici , G. Loupias , S. Rabii, S. Erwin, A. Rassat and C. Fabre, Europhys. Lett. **31**, 87 (1995).

[20] C. Metz, Th. Tschentscher, T. Sattler, K. Höppner, J. R. Schneider, K. Wittmaack, D. Frischke, and F. Bell, Phys. Rev. B **60**, 14049 (1999).

[21] A. Bansil, Z. Naturforsch. **48a**, 165 (1993).

[22] R. Benedek, R. Prasad, S. Manninen, B.K. Sharma, A. Bansil, and P.E. Mijnarends, Phys. Rev. B **32**, 7650 (1985).

[23] C. Sternemann, K. Hämäläinen, A. Kaprolat, A. Soininen, G. Döring, C.-C. Kao, S. Manninen, and W. Schülke, Phys. Rev. B **62**, R7687 (2000).

[24] S. Huotari, K. Hämäläinen, S. Manninen, S. Kaprzyk, A. Bansil, W. Caliebe, T. Buslaps, V. Honkimäki, and P. Suortti, Phys. Rev. B **62**, 7956 (2000).

[25] J. M. Luttinger, Phys. Rev.**119**, 1153 (1960).

[26] B. Lundqvist, Phys. Kondens. Mater.**7**, 117 (1968).

[27] C. Metz, Th. Tschentscher, P. Suortti, A.S. Kheifets, J.R. Lun, T. Sattler, J.R. Schneider, and F. Bell, Phys. Rev. B **59**, 10512 (1999).

[28] C. Sternemann, PhD Thesis, Fachbereich Physik, University of Dortmund, (2000).

[29] W. Schülke, G. Stutz, F. Wohlert, and A. Kaprolat, Phys. Rev. B **54**, 14381 (1996).

[30] P. Suortti, T. Buslaps, V. Honkimäki, C. Metz, A. Shukla, Th. Tschentscher, J. Kwiatkowska, F. Maniawski, A. Bansil, S. Kaprzyk, A. S. Kheifets, D. R. Lun, T. Sattler, J. R. Schneider, and F. Bell, J. Phys. Chem. Solids **61**, 397 (2000).

[31] T. K. Ng and B. Dabrowski, Phys. Rev. B **33**, 5358 (1986).

[32] F. Bell, Phys. Rev. B **67**, 155110 (2003).

[33] J. Fraxedas, J. Trodahl, S. Gopalan, L. Ley, and M. Cardona, Phys. Rev. B **41**, 10068 (1990).

[34] N.V. Smith, P. Thiry, and Y. Petroff, Phys. Rev. B **47**, 15476 (1993).

[35] A. vom Felde, J. Sprösser-Prou, and J. Fink, Phys. Rev. B **40**, 10181 (1989).

[36] K. Hämäläinen, D.P. Siddons, J.B. Hastings, and L.E. Berman, Phys. Rev. Lett. **67**, 2850 (1991).

[37] T.-C. Chiang, Chem. Phys. **251**, 133 (2000).

[38] E. Jensen, R.A. Bartynski, S.L. Hulbert, E.D. Johnson, and R. Garret, Phys. Rev. Lett. **62**, 71 (1989).

[39] W. Bothe and H. Geiger, Z. Phys. **32**, 639 (1925).

9 Theory of (e,2e) Spectroscopy from Ferromagnetic Surfaces

Roland Feder and Herbert Gollisch

We report some recent progress in the theory and understanding of electron pair emission from ferromagnetic crystalline surfaces, which occurs due to the collision of incident spin-polarized low-energy electrons with valence band electrons. A brief outline of the theoretical framework is followed by an analysis of transition matrix elements, which reveals spin and spatial selection rules for the (e,2e) reaction cross section. Valence electron states with spatial parts antisymmetric with respect to the reaction plane are not observable for any spin configuration. If there is a spatial symmetry plane normal to the reaction plane, spin-polarized primary electrons selectively probe equal- and opposite-spin valence electrons. Following an analytical specialization to cubic (110) surfaces, numerical results for Fe(110) are presented, which quantitatively illustrate a detailed interpretation of (e,2e) spectra in terms of the spin-split valence electron structure. surface state resonances are found to contribute very strongly. Calculated spectra are in good agreement with recent experimental data.

9.1 Introduction

The energy- and momentum-resolved investigation of pairs of time-correlated electrons, which are emitted from crystalline surfaces due to the collision of an incident low-energy (less than about 50 eV) electron with a valence electron has, over the past half-decade, strongly advanced both experimentally and theoretically, and is now established as a valuable method for studying electron collision dynamics, correlation effects and the electronic structure of surface systems (cf. Refs. [1–11] and references therein).

In particular, employing spin-polarized primary electrons in (e,2e) spectroscopy from ferromagnetic surfaces has been demonstrated to be useful for the study of (a) surface magnetism and (b) the exchange-induced spin dependence of electron scattering dynamics [4–6, 9, 10]. Evidently, a detailed understanding of the latter is required to extract information on surface magnetism from measured two-electron spectra, i.e. on the spin-split valence electron structure near the surface. In this chapter we show analytically that the consideration of symmetry properties of the spatial parts of the valence electron states contributes significantly to this understanding by virtue of spin- and spatial-symmetry-dependent selection rules. Furthermore, we present numerical results for ferromagnetic Fe(110), which, firstly, elucidate the relationship between the observable spin asymmetry of the reaction cross section and the valence electron spin polarization, secondly, reveal a strong surface sensitivity, and thirdly agree well with experimental data.

Correlation Spectroscopy of Surfaces, Thin Films, and Nanostructures. Edited by Jamal Berakdar, Jürgen Kirschner
Copyright © 2004 Wiley-VCH Verlag GmbH & Co. KGaA, Weinheim
ISBN: 3-527-40477-5

In Section 9.2 we briefly outline essential theoretical concepts and formulae. In Section 9.3 we deduce that spin-dependent (e,2e) reaction cross sections from ferromagnetic crystalline surface systems are subject to selection rules due to spatial symmetries of the valence electrons states involved. The results are illustrated analytically for cubic (110) surfaces. In Section 9.4 numerically calculated spin-dependent (e,2e) spectra for the Fe(110) surface are interpreted in terms of the underlying spin- and symmetry-resolved valence electron band structure and layer-resolved density of states, and compared with experimental data.

9.2 Concepts and Formalism

Since the present work is based on a theoretical framework for the (e,2e) process, which has already been presented in detail (cf. review article [11] and references therein), it may suffice here to outline the essentials required for the derivations and discussions in the subsequent sections.

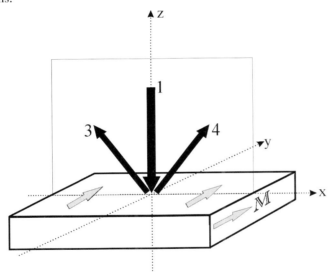

Figure 9.1: Schematic set-up of (e,2e) spectroscopy in coplanar symmetric geometry: upon impact of a primary electron (labeled as 1) on a magnetic surface and collision with a valence electron, two electrons (labeled as 3 and 4) exit into the vacuum region in the directions of two energy-analyzing detectors connected by coincidence circuitry.

A typical geometry of reflection-mode (e,2e) spectroscopy is sketched in Figure 9.1. A primary electron with energy E_1, surface-parallel momentum component $k_1^{\parallel}$ and spin orientation σ_1 (relative to the y-axis) collides with a valence electron, which is characterized by the corresponding set $(E_2, k_2^{\parallel}, \sigma_2)$, and two outgoing electrons with $(E_3, k_3^{\parallel}, \sigma_3)$ and $(E_4, k_4^{\parallel}, \sigma_4)$ are detected in coincidence. We note that even if there is no spin detection (as in present-day experiments), the spin labels σ_3 and σ_4 are still important and have finally to be summed over. The primary electron and the valence electron are described by solutions $\psi_1^{\sigma_1}(\mathbf{r})$ and $\psi_2^{n\sigma_2}(\mathbf{r})$

of an effective-potential one-electron Dirac equation, where the quantum numbers E_i and $k_i^\|$ are comprised in the respective subscripts 1 and 2, and the index n accounts for degeneracies of the valence electron state. If the Coulomb correlation between the two outgoing electrons is either neglected or incorporated in effective one-electron potentials (cf. Ref. [3]), these electrons are also described by one-electron four-spinors: $\psi_3^{\sigma_3}(\mathbf{r})$ and $\psi_4^{\sigma_4}(\mathbf{r})$. Since the energies and surface-parallel momenta of the primary and of the two detected electrons can be fixed, E_2 and $k_2^\|$ of the valence electron are determined by the conservation laws

$$E_2 = E_3 + E_4 - E_1 \qquad \text{and} \qquad k_2^\| = k_3^\| + k_4^\| - k_1^\| \tag{9.1}$$

The σ_i are, due to spin–orbit coupling, not good quantum numbers, but only labels for degenerate pairs of solutions, which for the primary and the detected electrons have to match to plane waves with definite spin projections σ_i in the vacuum region.

The fully spin-resolved (e,2e) reaction cross section ("intensity"), as obtained by first order perturbation theory in the screened Coulomb interaction $U(\mathbf{r}, \mathbf{r}')$, can then be expressed in terms of one-electron Green functions or equivalently in the following quasi-particle golden rule form

$$I_{\sigma_3,\sigma_4}^{\sigma_1}(1,3,4) = (k_3 k_4 / k_1) \sum_{E_2, \mathbf{k}_2^\|, n\sigma_2} |f_{\sigma_1, n\sigma_2, \sigma_3, \sigma_4} - g_{\sigma_1, n\sigma_2, \sigma_3, \sigma_4}|^2 \, \delta(E)\delta(\mathbf{k}^\|), \tag{9.2}$$

where the arguments $1, 3, 4$ abbreviate the sets $(E_i, \mathbf{k}_i^\|)$ with $i = 1, 3, 4$ and the δ-functions reflect Eq. (9.1); f and g are direct and exchange scattering amplitudes:

$$f_{\sigma_1, n\sigma_2, \sigma_3, \sigma_4} = \int \psi_3^{\sigma_3 *}(\mathbf{r}) \psi_4^{\sigma_4 *}(\mathbf{r}') U(\mathbf{r}, \mathbf{r}') \psi_1^{\sigma_1}(\mathbf{r}) \psi_2^{n\sigma_2}(\mathbf{r}') \, \mathrm{d}\mathbf{r}\mathrm{d}\mathbf{r}'; \tag{9.3}$$

the expression for g is the same except for $\mathbf{r}$ and $\mathbf{r}'$ interchanged in the first product term. Equation (9.2) relates to a "complete experiment" from a ferromagnetic surface with a spin-polarized source and detection of the spins of the two outgoing electrons. If the detectors do not resolve spin, one has to sum over σ_3 and σ_4. If the primary beam is unpolarized, one has to sum over $\sigma_1 = \pm$. We finally note that the above holds for arbitrary incidence and detection directions.

9.3   Spin and Spatial Selection Rules

Due to spin–orbit coupling, the direct and exchange amplitudes in Eq. (9.1) are, in principle, non-zero for all the 16 sets $(\sigma_1, \sigma_2, \sigma_3, \sigma_4)$. In practice however, there are many situations in which spin–orbit coupling is weak enough for results, which are derived in its absence, to be useful approximations. Of such nature are the spin and spatial (e,2e) selection rules, which we deduce in the following.

Without spin–orbit coupling the spinors $\psi_i^{\sigma_i}$ can be reduced to $\varphi_i^{\sigma_i}|\sigma_i\rangle$, where $|\sigma_i\rangle$ with $\sigma_i = \pm$ are the basis spinors $(1, 0)$ and $(0, 1)$, and $\varphi_i^{\sigma_i}$ are scalar wavefunctions. It is important to note that for ferromagnets the latter actually depend on the spin orientation σ_i. Nonvanishing amplitudes f and g are then easily seen to exist only for a few sets $(\sigma_1, \sigma_2, \sigma_3, \sigma_4)$, namely those for which the sum of the spin projections of the primary and the valence electron is conserved in the collision process.

If the spin of the primary electron is parallel to the spins of the valence electrons of spatial degeneracy type n, i.e. $\sigma_1 = \sigma_2 =: \sigma$, we have only

$$f_{\sigma,n\sigma,\sigma,\sigma} = \int \varphi_3^{\sigma*}(\boldsymbol{r})\varphi_4^{\sigma*}(\boldsymbol{r}')U(\boldsymbol{r},\boldsymbol{r}')\varphi_1^{\sigma}(\boldsymbol{r})\varphi_2^{n\sigma}(\boldsymbol{r}')\,\mathrm{d}\boldsymbol{r}\mathrm{d}\boldsymbol{r}' =: f_n^{\sigma\sigma} \tag{9.4}$$

and

$$g_{\sigma,n\sigma,\sigma,\sigma} = \int \varphi_3^{\sigma*}(\boldsymbol{r}')\varphi_4^{\sigma*}(\boldsymbol{r})U(\boldsymbol{r},\boldsymbol{r}')\varphi_1^{\sigma}(\boldsymbol{r})\varphi_2^{n\sigma}(\boldsymbol{r}')\,\mathrm{d}\boldsymbol{r}\mathrm{d}\boldsymbol{r}' =: g_n^{\sigma\sigma} \tag{9.5}$$

For primary spin σ, the cross section for detecting a parallel-spin σ electron pair is then

$$\begin{aligned}
I_{\sigma,\sigma}^{\sigma}(1,3,4) &= (k_3k_4/k_1) \sum_{E_2,\boldsymbol{k}_2^{\|},n} |f_n^{\sigma\sigma} - g_n^{\sigma\sigma}|^2 \,\delta(E)\delta(\boldsymbol{k}^{\|}) \\
&=: \sum_n I_n^{\sigma\sigma} =: I^{\sigma\sigma}.
\end{aligned} \tag{9.6}$$

The partial intensity $I_n^{\sigma\sigma}$ defined in Eq. (9.5) can be regarded as the scattering cross section of the spin σ primary electron with a spin σ valence electron of spatial degeneracy type n.

If the spins of the primary electron and the valence electrons are anti-parallel, to each other, i.e. $\sigma_2 = -\sigma_1 = -\sigma =: \bar{\sigma}$, the only non-vanishing scattering amplitudes are

$$f_{\sigma,n\bar{\sigma},\sigma,\bar{\sigma}} =: f_n^{\sigma\bar{\sigma}} \quad , \quad g_{\sigma,n\bar{\sigma},\bar{\sigma},\sigma} =: g_n^{\sigma\bar{\sigma}}, \tag{9.7}$$

which are given by matrix element expressions analogous to those in Eqs. (9.4) and (9.5). For primary spin σ, the cross section for detecting an antiparallel-spin pair is then

$$\begin{aligned}
I_{\sigma,\bar{\sigma}}^{\sigma} + I_{\bar{\sigma},\sigma}^{\sigma} &= (k_3k_4/k_1) \sum_{E_2,\boldsymbol{k}_2^{\|},n} \left(|f_n^{\sigma\bar{\sigma}}|^2 + |g_n^{\sigma\bar{\sigma}}|^2 \right) \delta(E)\delta(\boldsymbol{k}^{\|}) \\
&=: \sum_n I_n^{\sigma\bar{\sigma}} =: I^{\sigma\bar{\sigma}} .
\end{aligned} \tag{9.8}$$

The partial intensity $I_n^{\sigma\bar{\sigma}}$ defined in Eq. (9.8) is the scattering cross section of the spin σ primary electron with an opposite-spin $\bar{\sigma}$ valence electron of spatial degeneracy type n. In experiments with a polarized electron gun (spin $\sigma = \pm$) and without detection of the spins of the outgoing electrons, the observed scattering cross section is

$$I^{\sigma} := I^{\sigma\sigma} + I^{\sigma\bar{\sigma}} \quad \text{with} \quad \sigma = \pm . \tag{9.9}$$

In the general case of arbitrary directions of the incident electron and of the two outgoing electrons, all the cross sections $I_n^{\sigma\sigma}$ and $I_n^{\sigma\bar{\sigma}}$ (cf. Eqs. (9.6) and (9.8)) are usually non-zero. If by virtue of energy conservation (cf. Eq. (9.1)) the valence electron energy E_2 happens to lie in a majority spin energy gap (e.g. just below E_F for a strong ferromagnet like Ni), it is obvious that I_n^{++} and I_n^{-+} are zero.

In the following, we show that in more symmetrical geometries the spatial symmetries of the wavefunctions can cause some of the above partial cross sections to vanish, i.e. give rise to selection rules. We focus on co-planar setups, in which the incident and outgoing electrons move in a mirror plane of the semi-infinite crystal, which is perpendicular to the surface.

The wavefunctions $\varphi_1^{\sigma_1}$, $\varphi_3^{\sigma_3}$ and $\varphi_4^{\sigma_4}$, which are plane waves at the source and at the detectors, are then symmetric with respect to the mirror operation at that plane, while the valence electron wavefunctions $\varphi_2^{n\sigma_2}$ generally comprise one symmetric and one antisymmetric function. For the latter, which we characterize by the index $n = a$, the scattering amplitudes $f_a^{\sigma\sigma'}$ and $g_n^{\sigma\sigma'}$ as defined in Eqs. (9.4), (9.5) and (9.7) are identically zero. This is easily seen by e.g. choosing the mirror plane as (x, z) and applying the transformation $(y, y') \to (-y, -y')$ to the integrals in Eqs. (9.4), (9.5), which by virtue of $\varphi_2^{a\sigma_2}(x, -y, z) = -\varphi_2^{a\sigma_2}(x, y, z)$ yields $f_a^{\sigma\sigma'} = -f_a^{\sigma\sigma'}$ and $g_a^{\sigma\sigma'} = -g_a^{\sigma\sigma'}$. Consequently, the contributions to the cross sections $I^{\sigma\sigma}$ (Eq. (9.6)) and $I^{\sigma\bar{\sigma}}$ (Eq. (9.8)) are identically zero. We thus have the selection rule that valence electron states, which are antisymmetric with respect to the reaction plane, do not manifest themselves in (e,2e) reaction cross sections.

A second selection rule can be derived for coplanar geometry in the more special case of normal incidence of the primary beam and of equal energies ($E_3 = E_4$) and oppositely equal parallel-momenta of the detected electrons ($\mathbf{k}_3^{\|} = -\mathbf{k}_4^{\|}$). From parallel-momentum conservation (cf. Eq. (9.2)) this implies that one selects valence electrons states with $\mathbf{k}_2^{\|} = 0$, i.e. at the center of the surface Brillouin zone. Let (x, z) be the reaction plane and (y, z) a mirror plane of the semi-infinite crystalline system with the surface (clean or covered with an ultrathin film) in the (x, y) plane. The valence electron wavefunctions are then either symmetric or antisymmetric with respect to the (y, z) plane. For parallel spins $\sigma_3 = \sigma_4 =: \sigma$, the wavefunctions of the two outgoing electrons are related as $\varphi_4^{\sigma}(x, y, z) = \varphi_3^{\sigma}(-x, y, z)$. Applying the transformation $(x, x') \to (-x, -x')$ to the integral in Eq. (9.5) and comparing with Eq. (9.4) reveals that $g_n^{\sigma\sigma} = f_n^{\sigma\sigma}$ if $\varphi_2^{n\sigma}$ is symmetric, and $g_n^{\sigma\sigma} = -f_n^{\sigma\sigma}$ if $\varphi_2^{n\sigma}$ is antisymmetric. The cross section contribution $I_n^{\sigma\sigma}$ (Eq. (9.6)) for the production of parallel-spin pairs therefore vanishes in the former case, whereas in the latter it is non-zero and determined by $4|f_n^{\sigma\sigma}|^2$. $I^{\sigma\sigma}$ (Eq. (9.6)), the scattering cross section of a spin σ primary electron with a spin σ valence electron, thus selectively reflects collisions involving valence electron states, which are antisymmetric with respect to a mirror plane perpendicular to the reaction plane.

In the following we apply the above-derived selection rules analytically to ferromagnetic crystalline surface systems, which in the non-magnetic limit have at least a mirror symmetry plane normal to the surface. For coplanar (e,2e) set-ups with the reaction plane parallel to this mirror plane, parallel-momentum conservation (cf. Eq. 9.1) implies that the relevant valence electrons then have $\mathbf{k}_2^{\|}$ in this plane and wavefunctions $\varphi_2^{n\sigma}(\mathbf{r})$ (with $\sigma = \pm$) with $n = 1$ and 2, which under the mirror operation are symmetric and antisymmetric, respectively. According to our first selection rule, it is only the symmetric ones that can manifest themselves in (e,2e) spectra.

For surface systems, which have a second mirror plane and a two-fold rotation axis normal to the surface (spatial symmetry group 2 *mm*), more detailed information becomes available by choosing symmetric coplanar set-ups with normal incidence of the primary beam and

Table 9.1: (e,2e) selection rules in coplanar symmetric equal-energy set-up from ferromagnetic crystalline surface systems with a two-fold rotation axis and two mirror planes normal to the surface (spatial symmetry group 2 *mm*). For valence states with spatial symmetry types $\Sigma_1, \Sigma_2, \Sigma_3, \Sigma_4$ the first two rows indicate the even and odd mirror symmetry (+ and −) of the state with respect to the (x, z) and (y, z) plane, where z is normal to the surface. For bcc and fcc(110) surfaces, the in-surface-plane coordinates x and y are chosen along [1 −1 0] and [001], respectively. The subsequent rows give the contributions to the anti-parallel-spin cross section $I^{\sigma\bar\sigma}$ (Eq. (9.8)) and the parallel-spin cross section $I^{\sigma\sigma}$ (Eq. (9.6)), for the reaction plane chosen firstly as (x, z) and secondly as (y, z).

	Valence Electron Mirror Symmetry Plane		(e,2e) in (x, z) Plane Contributions to		(e,2e) in (y, z) Plane Contributions to	
	(x, z)	(y, z)	$I^{\sigma\bar\sigma}$	$I^{\sigma\sigma}$	$I^{\sigma\bar\sigma}$	$I^{\sigma\sigma}$
Σ_1	+	+	$2\|f_1^{\sigma\bar\sigma}\|^2$	0	$2\|f_1^{\sigma\bar\sigma}\|^2$	0
Σ_2	−	−	0	0	0	0
Σ_3	−	+	0	0	$2\|f_3^{\sigma\bar\sigma}\|^2$	$4\|f_3^{\sigma\sigma}\|^2$
Σ_4	+	−	$2\|f_4^{\sigma\bar\sigma}\|^2$	$4\|f_4^{\sigma\sigma}\|^2$	0	0

equal energies of the two detected electrons (cf. Figure 9.1). The relevant valence electrons then have $k_2^\parallel = 0$ and their (crystal half-space plus vacuum-half-space) wavefunctions $\varphi_2^{n\sigma}(r)$ (with $\sigma = \pm$) can be classified according to the spatial symmetry types Σ_n with $n = 1, 2, 3, 4$. For a coordinate system with z along the surface normal and x and y in the surface plane the two surface-perpendicular mirror planes are (x, z) and (y, z). The valence electron wavefunctions then behave as follows under reflection at the two mirror planes: $\varphi_2^{1\sigma}$ is symmetric with respect to both, $\varphi_2^{2\sigma}$ is antisymmetric with respect to both; $\varphi_2^{3\sigma}$ is antisymmetric with respect to (x, z) and symmetric with respect to (y, z), and $\varphi_2^{4\sigma}$ is symmetric with respect to (x, z) and antisymmetric with respect to (y, z).

As the (e,2e) reaction plane we choose firstly the (x, z) and secondly the (y, z) plane. We then easily obtain the (e,2e) intensity contributions $I_n^{\sigma\bar\sigma}$ (cf. Eq. (9.6)) and $I_n^{\sigma\sigma}$ (cf. Eq. (9.8)), which are summarized in Table 9.1.

Valence electron states of Σ_1 symmetry contribute to $I^{\sigma\bar\sigma}$, but not to $I^{\sigma\sigma}$ due to their symmetry with respect to the (y, z) plane and consequently $g_1 = f_1$ (second selection rule). Valence electron states of Σ_2 symmetry cannot be "seen" in the (e,2e) spectra for both reaction planes. For the (x, z) reaction plane, there is furthermore no contribution from Σ_3 states on the grounds of our first selection rule, but Σ_4 states do contribute; most notably also to $I^{\sigma\sigma}$, since their antisymmetry with respect to reflection at the (y, z) plane entails $g_4 = -f_4$ (see Eq. (9.6)). The scattering cross section of a spin σ primary electron with a spin σ valence electron stems thus exclusively from valence electrons of Σ_4 symmetry. For the (x, z) reaction plane, the results are analogous, with Σ_3 and Σ_4 interchanged.

9.4   Numerical Results for Fe(110)

We have performed numerical calculations of the spin- and layer-resolved density of states and of spin-resolved (e,2e) spectra for the ferromagnetic Fe(110) surface by means of a layer-KKR code [12], in which an equivalent of the above golden rule expression Eq. (9.2) in terms of the valence electron spectral function has been implemented. The real part of the effective one-electron potential was obtained from a nine-layer-film self-consistent GGA-FLAPW [13] charge density. As the imaginary part of the potential we chose, guided by LEED experience, $-0.07(E_i - E_F)$ eV with $i = 1$ for the incident and $i = 3, 4$ for the emitted electrons. For the valence electrons, we firstly took a rather small value -0.05 eV in order to make the symmetry analysis and interpretation more transparent. and secondly the more realistic form $0.25(E_2 - E_F) - 0.15$ eV in view of comparing with experimental data.

As a basis for the interpretation of (e,2e) spectra we show in Figure 9.2 the spin-resolved bulk band structure of Fe along $\Gamma(\Sigma)N$ and the spin- and spatial-symmetry-resolved bulk and surfaces layer densities of states (LDOS) of the Fe(110) surface for $k^\parallel = 0$, i.e. at the center of the surface Brillouin zone. For each symmetry type Σ_i the bulk LDOS features are seen to be associated with the bands in the usual way. The surface LDOS (topmost atomic layer) is significantly different. In particular, we notice a distinct minority spin surface resonance feature of Σ_1 symmetry at -0.25 eV and majority spin surface state or resonance peaks of Σ_1 and Σ_4 symmetry at lower energies. These features have already been observed by spin- and angle-resolved photoemission spectroscopy and discussed in detail in conjunction with self-consistent slab calculations [14].

In Figure 9.3a we present, for symmetric coplanar geometry with the reaction plane chosen as the (x, z) mirror plane (with x along the $[1, -1, 0]$ direction) (cf. Figure 9.1), a typical set of equal-energy-sharing (e,2e) cross sections from Fe(110) as functions of the valence electron energy relative to the Fermi energy. The solid and dashed spectra in all panels were obtained for spin-up and spin-down primary electrons, respectively. In the upper two panels, they have been decomposed according to the spins of the two emitted electrons (cf. Eqs. (9.6) and (9.8)). The topmost spectrum is due to the collision of spin-up primary electrons with spin-up valence electrons of Σ_4 symmetry. Its peak is seen to be at the same energy as the Σ_4 majority-spin surface LDOS feature in Figure 9.2. The spectra in the second panel are due to collisions of spin-up and spin-down primary electrons with minority and majority spin valence electrons, respectively. Comparison with the valence electron LDOS in Figure 9.2 reveals their detailed origin as indicated by the symbols 1^+, 1^- and 4^+ above the respective peaks. The locations of the peaks are seen to coincide with the corresponding features of the surface LDOS rather than those of the bulk LDOS. This demonstrates, in line with earlier findings [7], that (e,2e) spectroscopy is extremely surface-sensitive. Valence electrons of Σ_2 and Σ_3 symmetry do not manifest themselves in the calculated (e,2e) spectra, in accordance with our analytical selection rule results (cf. Table 9.1). Summation over the solid (dashed) spectra in the upper two panels yields the spectrum I^+ (I^-) produced by spin-up (spin-down) primary electrons without spin analysis of the emitted electron pair (third panel in Figure 9.3a). We would like to emphasize that I^+ reflects both minority and majority spin valence electron states.

Analogous results obtained for the (y, z) reaction plane are shown in Figure 9.3b. In this case, Σ_2 and Σ_4 valence electrons do not contribute, whereas the majority spin Σ_3 surface density of states (cf. Figure 9.2a) is seen to be reflected in the cross section for collisions

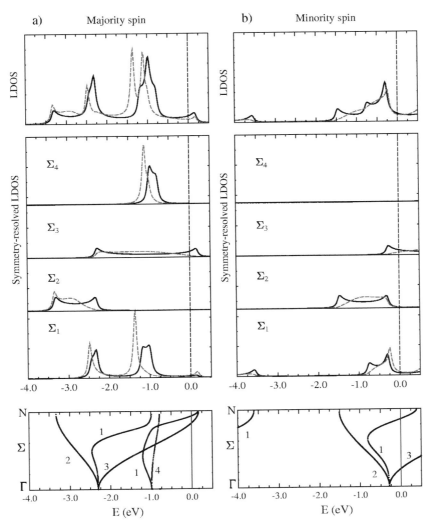

Figure 9.2: Valence electron states (without spin–orbit coupling) of ferromagnetic Fe(110):
(a) For majority spin: Bottom panel: Bulk band structure along $\Gamma(\Sigma)N$ with the symbols 1 to 4
indicating the spatial symmetry types $\Sigma_1, \Sigma_2, \Sigma_3, \Sigma_4$ of the individual bands. Top panels: Bulk
(solid lines) and surface (broken lines) layer-resolved densities of states (LDOS) for the four
symmetry types Σ_i and their sum (in the uppermost panel). (b) As (a), but for minority spin.
(The Σ_4-LDOS is negligible in the present plotting range.).

with spin-up primary electrons. The Σ_1-derived features in the second panel are at the same
energies as in the case of the (x, z) reaction plane, but their relative heights are significantly
different. The latter result is due to the fact that the off-normal time-reversed LEED states ψ_3
and ψ_4, and consequently the transition matrix elements, are quite different in the two cases.

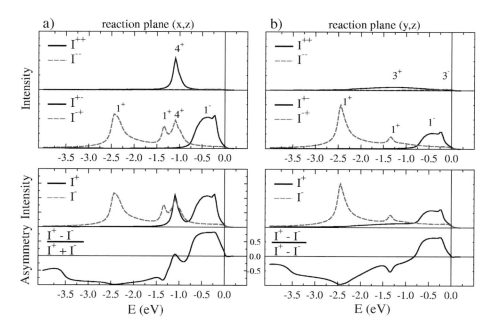

Figure 9.3: (a) Calculated (e,2e) spectra from the Fe(110) surface in symmetric coplanar geometry with reaction plane (x, z) (intersecting the surface along $[1 -1\ 0]$) for normally incident
primary electrons with energy $E_1 = 22.7$ eV and spin polarization along the majority (solid
lines) and minority spin (dashed lines) directions. The two emitted electrons have equal energies $E_3 = E_4$ and propagate at fixed polar angles $\theta_3 = \theta_4 = 40°$ relative to the surface normal
(z-axis). The intensities $I^{\sigma\sigma}$ (Eq. (9.6)), $I^{\sigma\bar{\sigma}}$ (Eq. (9.8)) and their sums $I^{\sigma} = I^{\sigma\sigma} + I^{\sigma\bar{\sigma}}$ are
thus functions of $E_3 = E_4$ and can, because of energy conservation $E_2 = 2E_3 - E_1$, also
be regarded as functions of the valence electron energy $E_2 =: E$, which we take (relative to
the Fermi energy) as the abscissa. In the upper two panels, the symbols 1^+, 1^- and 4^+ above
individual peaks indicate their origin from majority ($+$) and minority ($-$) valence electron states
of Σ_1 and Σ_4 symmetry. (b): Same as (a) but for reaction plane (y, z) (intersecting the surface
along $[0\ 0\ 1]$) and the symbols 3^+ and 3^- indicating that the respective peaks originate from Σ_3
majority and minority spin valence electrons.

The (e,2e) spectra in Figure 9.3 have been calculated assuming very long valence hole
life-times. For more realistic hole life-times, described by a larger and energy-dependent
imaginary potential part $0.25(E - E_F) - 0.15$ eV, we show in Figure 9.4 the corresponding
spectra I^+ and I^- for the (x, z) reaction plane. Although substantially broadened, all the
features, which were observed in Figure 9.3a, are still present. In particular, in the I^+ spectrum
the shoulder around -0.25 eV stems from the minority-spin Σ_1 surface resonance and the
hump around -1 eV is due to collisions with majority-spin Σ_4 valence electrons. If the degree
of spin polarization of the primary electron beam is only 50% (instead of the 100% assumed
in the above), the calculated I^+ and I^- spectra are those shown in the central panel. The
agreement with experimental data [15] is very good between the Fermi energy and -1.2 eV.

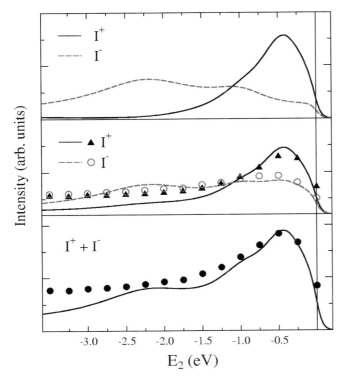

Figure 9.4: Spin-dependent equal-energy-sharing spectra I^+ and I^- (as functions of the valence electron energy E_2 (relative to E_F)) from ferromagnetic Fe(110) (with magnetization M parallel to the surface along $[1,0,0]$ (y axis)) in coplanar symmetric geometry (reaction plane (x,z)) with normal incidence of 22.6 eV polarized primary electrons and the two detectors at polar angles $\theta_3 = \theta_4 = 40°$. Our theoretical results (solid and broken curves), obtained for primary electron degree of spin polarization 100% (topmost panel) and 50% (central and bottom panels) are compared to experimental data (triangles and circles) from Kirschner's group [15]. While $E_3 - E_4$ is exactly zero for the calculated spectra, it covers a range between -0.1 and $+0.1$ $(E_3 + E_4)$ (in eV) in the experimental data.

At lower energies the experimental spin splitting has the same sign as the theoretical one, but is much smaller. Amongst the various factors contributing to the latter discrepancy, an important one is presumably that in the calculations we take into account only a single electron–electron collision, whereas the experimental detection at lower energies also includes electrons, which have undergone further energy loss processes (e.g. electron–hole or plasmon excitation). Since these electrons stem predominantly from a first collision of a spin-up primary electron with a spin-down valence electron, they mainly add to the measured I^+ spectrum below -1.2 eV. And indeed, if in this energy range one subtracts some background from the experimental I^+ spectrum, agreement with theory does improve.

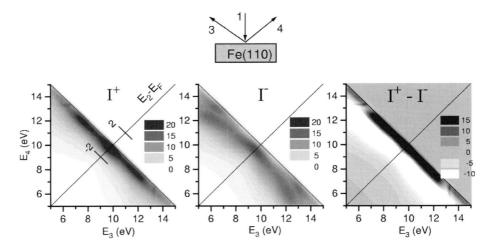

Figure 9.5: Spin-polarized (e,2e) from ferromagnetic Fe(110) with setup as described in the caption to Figure 9.4. The contour plots show calculated reaction cross sections I^+ and I^- and their difference, as functions of E_3 and E_4, for 100% spin polarization of the primary electrons.

The above-shown (e,2e) spectra have been obtained for outgoing electron energies $E_3 = E_4$ and are due to collisions with valence electrons with $k_2^{\parallel} = 0$. Valence electrons with $k_2^{\parallel} \neq 0$ can – in the coplanar symmetric geometry – be probed by choosing $E_3 \neq E_4$ subject to the energy conservation condition Eq. (9.1). For fixed primary energy E_1, the spin-dependent (e,2e) cross sections are then functions $I^{\sigma}(E_3, E_4)$ and can be represented by contour plots in the (E_3, E_4) plane. By virtue of the conservation conditions Eq. (9.1) a pair $(E_2, k_2^{\parallel})$ is uniquely associated with each pair (E_3, E_4). A set of such contour plots, which we have calculated for Fe(110), is shown in Figure 9.5. The diagonal line, along which valence electrons with $k_2^{\parallel} = 0$ are probed, can be regarded as the valence electron E_2 axis. The counter-diagonal line marks the Fermi energy E_F and can be viewed as a (non-linear) $k_2^{\parallel}$ axis. As is most clearly seen in the difference spectrum, I^+ dominates in a stripe between E_F and about -1 eV, whereas I^- prevails below about -1 eV. In a qualitative and overall manner, I^+ (I^-) thus reflects minority-spin (majority-spin) valence electrons. This correspondence is however not quantitative for several reasons: (a) valence states, which are antisymmetric to the reaction plane, are not represented at all, (b) opposite-spin electrons increasingly contribute to the partial intensities as one goes away from the diagonal, and (c) matrix element effects are of considerable importance. Comparing the theoretical results in Figure 9.5 with recently published experimental data [10] we notice a good overall agreement from E_F down to about -2 eV. For lower energies, the experimental data increasingly contain contributions due to multiple collision processes, which are not taken into account in our theory.

In order to assess the surface sensitivity of the spectra, we performed analogous calculations, in which we suppressed surface state and resonance features by replacing the realistic continuous surface potential barrier by a non-reflecting one. Both I^+ and I^- turned out to be significantly weakened. The same was observed experimentally, when the surface was contaminated by oxygen [15]. We recall in this context, that detailed ground state electronic structure calculations revealed surface states and resonances over wide ranges of the surface Brillouin zone [14]. Our results thus corroborate that spin-dependent (e,2e) spectroscopy experiments in conjunction with theory provide a powerful tool for studying surface magnetism.

Acknowledgment

We would like to thank J. Kirschner and A. Morozov (Max-Planck-Institut für Mikrostrukturphysik, Halle) for making available recent experimental data and discussing them with us.

References

[1] R. Feder, H. Gollisch, D. Meinert, T. Scheunemann, O.M. Artamonov, S. Samarin and J. Kirschner, Phys. Rev. B **58**, 16418 (1998).

[2] H. Gollisch, T. Scheunemann, and R. Feder, J. Phys.: Condens. Matter **11**, 9555 (1999).

[3] J. Berakdar, H. Gollisch and R. Feder, Solid State Commun. **112**, 587 (1999).

[4] J. Berakdar, Phys. Rev. Lett. **83**, 5150 (1999).

[5] S.N. Samarin, J. Berakdar, O. Artamonov and J. Kirschner, Phys. Rev. Lett. **85**, 1746 (2000).

[6] S.N. Samarin, O. Artamonov, J. Berakdar, A. Morozov and J. Kirschner, Surf. Sci. **482**, 1015 (2001).

[7] H. Gollisch, T. Scheunemann, and R. Feder, Solid State Commun. **117**, 691 (2001).

[8] R. Feder and H. Gollisch, Solid State Commun. **117**, 691 (2001).

[9] R. Feder, H. Gollisch, T. Scheunemann, J. Berakdar and J. Henk, in J. Berakdar and J. Kirschner (Eds), *Many-Particle Spectroscopy of Atoms, Molecules and Surfaces*, Kluwer Academic / Plenum, New York - London (2001).

[10] A. Morozov, J. Berakdar, S.N. Samarin, F.U. Hillebrecht and J. Kirschner, Phys. Rev.B. **65**, 104425 (2002).

[11] R. Feder and H. Gollisch in W. Schattke and M.A. Van Hove (Eds), *Solid State Photoemission and Related Methods*, Wiley-VCH, Weinheim (2003).

[12] S.V. Halilov, E. Tamura, H. Gollisch, D. Meinert and R. Feder, J. Phys. Condens. Matter **5**, 3859 (1993).

[13] P. Blaha, K. Schwarz and J. Luiz, computer code WIEN97, Vienna University of Technology, 1997. Improved and updated version of the original copyrighted WIEN-code, which was published by P. Blaha, K. Schwarz, P. Sorantin and S. B. Trickey in Comput. Phys. Commun. **59**, 399 (1990).

[14] H.J. Kim, E Vescovo, S. Heinze and S. Bluegel, Surf. Sci. **478**, 193 (2001).

[15] J. Kirschner and A. Morozov, personal communication.

10 Ab-initio Calculations of Charge Exchange in Ion–surface Collisions: An Embedded–cluster Approach

Ludger Wirtz, Michal Dallos, Hans Lischka, and Joachim Burgdörfer

We discuss the feasibility of the embedded cluster approach for ab-initio calculations of charge exchange between ions and a LiF surface. We show that the discrete density of valence states in embedded clusters converges towards the continuum limit of the density of states in the valence band of an infinitely extended LiF surface. Screening of the holes that are left in the surface after electron transfer to the projectile plays an important role for the correct level ordering in the calculation of potential energy surfaces. We discuss to what extent the hole screening is taken into account by different levels of approximations which are customarily employed in quantum chemistry. The central result of this chapter is the convergence of potential energy curves with respect to cluster size: Out of the increasing number of potential energy curves (converging towards a continuum for infinite cluster size), only a small number of states effectively interacts with the capture level of the projectile and determines the charge transfer efficiency.

10.1 Introduction

Charge exchange plays a major role in the collision of ions with surfaces. An observable readily accessible in experiments is the final charge state of an ion after scattering at the surface. Also for the description of other experimentally observable quantities, a detailed knowledge of the charge transfer dynamics is desirable: For example, the time-dependent charge state of the projectile determines the interaction potential with the surface and thereby influences the projectile trajectory, i.e. the energy and the angle of backscattered ions. Furthermore, in insulators with strong electron–phonon coupling, electron transfer to the projectile can lead to formation of self-trapped defects (electron holes, excitons) which, in turn, can result in the ablation of secondary particles from the surface [1].

Despite the importance of charge transfer for virtually all phenomena involving ion–surface collisions, an accurate ab-initio treatment is still missing. This is, of course, due to the complexity of the problem. In particular, in the case of insulator surfaces, where the description of the surface in terms of the jellium model (assuming a homogeneous positive background charge instead of localized atomic cores) is not suitable, the dynamics of a many-nuclei and many-electron system must be treated explicitly. The interaction of (discrete) projectile states with the continuum of states in the surface valence band entails both the properties of the infinitely extended surface and the localized projectile state. The former is usually achieved by using Bloch wavefunctions and describing the system in a supercell (consisting

Correlation Spectroscopy of Surfaces, Thin Films, and Nanostructures. Edited by Jamal Berakdar, Jürgen Kirschner
Copyright © 2004 Wiley-VCH Verlag GmbH & Co. KGaA, Weinheim
ISBN: 3-527-40477-5

of a two-dimensional unit cell parallel to the surface and a large slab of bulk and vacuum in a perpendicular direction). In contrast, the localized interaction of the projectile ion with one or several atoms of the surface is more appropriately described by the methods of ion–atom/ion–molecule collision. There are two possibilities to combine both approaches: One possibility would be to treat the ion–surface collision in a supercell. However, apart of the exceedingly large size of the supercell, additional difficulties would arise due to the positive net charge of the projectile. The long-range Coulomb potential of the periodically repeated positive projectile would have to be artificially screened in order not to affect neighboring unit cells and a negative background charge would have to be introduced in order to render the supercell neutral. Alternatively, in the approach pursued in the following we choose the second option which is the calculation of a projectile-collision with a cluster of surface ions embedded into a large array of point charges that represents the residual (infinitely extended) surface and bulk.

In order to render the embedded-cluster approach valuable for the description of the interaction of the projectile with an infinitely extended surface, several criteria have to be met:

1. The (discrete) density of states of the embedded cluster should – in the limit of large cluster size – approach the continuum limit of the density of states of the infinite system.

2. The ionization energy of the embedded cluster should agree with the workfunction of the surface. This point is important for the proper energetic ordering of the projectile state relative to the valence band. This is a highly non-trivial requirement as the Hartree–Fock theory is well-known to overestimate the band gap of insulators by up to several eV while density functional theory (DFT) underestimates it by about the same amount [2]. The proper treatment of electron correlations is therefore indispensable. The main effect of correlation in the current context is the screening of the hole that is left behind in the surface when an electron leaves the surface. This screening, i.e., the polarization of the environment, reduces the interaction of the hole with the emitted electron and reduces the ionization energy by up to several eV with respect to the value obtained by the Hartree–Fock approximation.

3. The potential energy curves that determine the charge exchange between projectile and surface must have converged as a function of cluster size.

Requirement (1) is analyzed in Section 10.2 where we compare the density of states (DOS) in the limit of large cluster size with the DOS obtained by a supercell calculation. Fulfilling the second criterion requires obviously a methodology that goes well beyond both Hartree–Fock theory and ground-state DFT. In Section 10.3, we summarize our approach [3] which is based on the quantum chemistry code COLUMBUS [4]. We use the multi-configuration self-consistent field (MCSCF) and multi-reference configuration interaction (MR-CI) approaches also taking into account size-consistency corrections. In Section 10.4 we present calculations of a H^+ ion impinging on embedded surface clusters of increasing size. We show that condition (3), i.e., the convergence with respect to cluster size is, indeed, fulfilled. Increasing the cluster size adds additional levels which, however, do not effectively interact with the projectile level. The chapter closes with remarks concerning the quantitative accuracy of our method and possible improvements.

In our calculations we use LiF as a surface material. LiF is a prototype of a wide band gap (14 eV) insulator and is also used in many experiments because it is a material with strong electron–phonon coupling and displays the effect of potential sputtering under the impact of slow ions [5, 6].

10.2   Convergence of the Density of States as a Function of Cluster Size

We present in this section a systematic study of the convergence of the density of states (DOS) of the valence electrons in a (bulk) embedded cluster of LiF towards the DOS of the infinitely extended system. We have performed Hartree–Fock (also referred to as self-consistent field, SCF) calculations for cubic clusters containing from 1^3 (single embedded F^-) up to 5^3 atoms[1]. In order to simulate the Madelung potential of the residual infinite crystal, the active clusters are embedded in a cubic array of negative and positive point charges at the positions of the F^- and Li^+ ions, respectively[2].

Table 10.1: Clusters used in the convergence study of the valence DOS. For clusters with odd ion number we calculate both the case with a fluorine in the center and that with a lithium in the center.

Size	Cluster	Cluster with Coordinated Ions
1^3	F	Li_5F
2^3	Li_4F_4	$Li_{16}F_4$
3^3	$Li_{14}F_{13}/Li_{13}F_{14}$	$Li_{38}F_{13}/Li_{43}F_{14}$
4^3	$Li_{32}F_{32}$	$Li_{80}F_{32}$
5^3	$Li_{63}F_{62}/Li_{62}F_{63}$	$Li_{135}F_{62}/Li_{146}F_{63}$

Table 10.1 shows the clusters for which we have performed calculations. The positive point charges at the border between the active cluster and the surrounding point charges are replaced by active Li^+ ions such that all active fluorines are fully coordinated by six lithium ions. This prevents an artificial distortion of the electron density at the border of the active cluster due to missing Pauli repulsion from the positive point charges.

In Figure 10.1(a) we present the orbital energies of the highest and lowest F_{2p}-like orbitals (valence orbitals) of the clusters listed in Table 10.1. The three F_{2p} orbitals of the embedded $Li_5^+F^-$ cluster are degenerate at an orbital energy of -15.5 eV. The transition to the next larger cluster with four F^- ions introduces a splitting of almost 1.5 eV. With increasing size, the band width increases more slowly and converges towards a value of 3.5 eV as can be seen

---

[1] The calculations were performed with the quantum chemistry code TURBOMOLE [7]. Since we deal with very large clusters, we use the pseudo-basis of Ref. [8] where the 1s core electrons are replaced by pseudo-potentials. For the Li^+ ions, we furthermore truncate the p-functions from the pseudo-basis. We have checked for small clusters that the influence of this truncation on the difference between valence electron levels is negligible ($< 10^{-3}$ eV).

[2] For active clusters with an even number of constituents, the number of point charges is such that the complete system (active cluster plus surrounding charges) contains 4096 ($= 16^3$) ions. For active clusters with uneven number of constituents, we use a total number of 3375 ($= 15^3$) ions. In this latter case, the system is charged which leads to a shift of the orbital energies of the active cluster. In order to enforce charge neutrality, we reduce/increment the charges at the eight corners of the cubic array of point charges by 1/8.

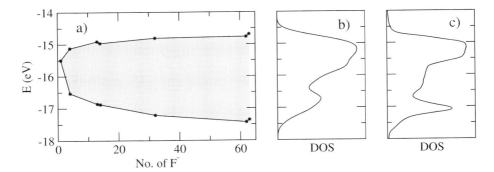

Figure 10.1: a) Orbital energy of highest and lowest F_{2p} like orbital as a function of the number of F^- contained in the cluster. b) Average density of F_{2p} states in the $Li_{135}^+F_{62}^-$ and $Li_{146}^+F_{63}^-$ clusters. Each discrete state is represented by a Gaussian peak with a full width at half maximum of 0.4 eV. c) DOS of infinite LiF calculated with DFT-LDA in a periodic supercell approach.

by plotting the band width as a function of the inverse linear dimension of the cluster [3]. This value agrees with the value obtained by photoelectron spectroscopy [9] and with quasi-particle band-structure calculations [10]. Figure 10.1(b) shows the average of the (orbital energy) DOS of the embedded $Li_{135}^+F_{62}^-$ and $Li_{146}^+F_{63}^-$ clusters[3]. In addition to the main peak at -15.2 eV, the DOS displays a side peak at -16.8 eV. This secondary peak is also seen in the experiment [9]. We compare the cluster DOS of Figure 10.1(b) with the DOS of an infinitely extended LiF crystal in Figure 10.1(c) for which the calculation[4] has been performed using density functional theory (DFT) in the local density approximation (LDA). The good agreement leads us to conclude that, in the limit of large clusters, the embedded cluster approach does indeed reproduce bulk quantities.

10.3   Going beyond Hartree–Fock

According to Koopmans' theorem, the energy of the highest occupied molecular orbital (HOMO) should be a good approximation to the ionization energy of the system, i.e. the work function of the infinitely extended surface. The experimental work function has a value of about $W_{LiF} = 12.3$ eV [12] which is smaller by more than 2 eV than the value of the HOMO energy (Figure 10.1) extrapolated to infinite cluster size. Increasing the basis set would lower the orbital energies by an additional eV upon convergence with respect to basis set size and render the discrepancy between the experimental value and the HOMO energy even larger. This discrepancy is not a failure of the embedded cluster approach but a failure of the Hartree–Fock method and is in line with the overestimate of the band gap of insulators

---

[3] In order to make the transition from discrete levels (corresponding to δ-peaks) to a continuous DOS, we introduce a Gaussian broadening of 0.4 eV (full width at half maximum) for each level.

[4] The calculation has been performed with the code ABINIT [11]. Wavefunctions are expanded in plane waves with an energy cutoff at 40 Hartree. Core electrons are described by Trouiller–Martins pseudopotentials. We have shifted the energy scale such that the upper edge of the bulk DOS and the embedded cluster DOS coincide.

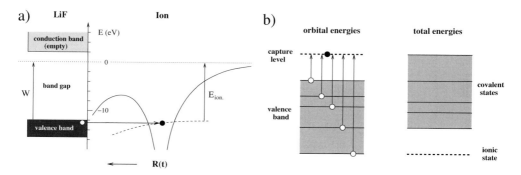

Figure 10.2: a) Orbital energy picture for the charge exchange between projectile ion and surface: schematic picture of band structure of LiF and of the capture level in the Coulombic potential of the ion core. As the projectile approaches the surface, the capture level is shifted due to electron–hole interaction and the dielectric response of the surface. b) Comparison of the orbital energy and total energy picture for the case where the capture level is higher than the valence band. Energy is required to transfer an electron from the band states to the capture level. Therefore, in the total energy picture, the covalent states (hole in the band + neutral projectile) are higher in energy than the ionic state (positive projectile + neutral band).

by up to several eV [2, 13]. The underlying reason is the neglect of screening of a hole left behind in the surface after ionization. The screening, i.e., the polarization of the environment, reduces the interaction of the hole with the emitted electron and thereby lowers the ionization energy. Screening cannot be properly described in a quasi-one particle picture underlying the Hartree–Fock approximation. Instead, methods that go beyond Hartree–Fock and include many-body effects are required.

Beyond the level of a mean-field approximation, the picture of orbital energies which is frequently invoked in the description of charge exchange phenomena and which is intrinsically connected to the one-particle picture, loses its meaning. Let us consider, e.g., the case of a H^+ ion colliding with a LiF surface as depicted schematically in Figure 10.2(a). The valence band extends from $-W_{LiF} = -12.3$ eV down to about -15.8 eV. The lowest projectile level into which an electron can be captured is the ground state of hydrogen at -13.6 eV which is thus, at a large projectile–surface distance, energetically positioned inside the valence band. Therefore, a strong interaction of this "capture level" with the valence band of LiF is expected facilitating charge exchange at small ion–surface distances. However, since on the Hartree–Fock level the top of the valence band lies too low by $2-3$ eV, the capture level lies well above the valence band during the approach to the LiF surface and charge exchange is suppressed at this level of description [3]. The proper level ordering in the combined ion–surface system is thus directly determined by the value of the work function of the system and requires the use of methods going beyond the Hartree Fock approximation. In turn, the concept of orbital energies which is related to the effective one-particle character of Hartree–Fock theory (or similarly, DFT) is no longer well-defined. The appropriate framework to describe charge exchange is therefore the calculation of *total* potential energy surfaces along ionic trajectories, i.e., the energies of ground and excited states of the system comprising the embedded cluster and the projectile ion with the position of the projectile as an adiabatic parameter. One of these N-electron states, the "ionic" state, corresponds at large distances $R \to \infty$ to the neutral surface with the positive ion in front while all the other states correspond to the projectile in a neutralized state with a hole left behind in the surface (see Figure 10.2(b)). Inclusion of correlation effects allows for a proper calculation of the work function of LiF and leads to a correct ordering of the *total* energies of ionic and covalent states of the combined projectile–surface system.

Our numerical approach has been described in detail in Ref. [3]. Here, we just give a brief summary of the method. We employ the quantum chemistry code COLUMBUS which is specifically designed for the calculation of ground and excited states through multi-reference and multi-configuration methods. The first step beyond Hartree–Fock or the single Slater-determinant self consistent field (SCF) method is the multi-configuration self-consistent field (MCSCF) method [14] which expands the many-electron wavefunction in different *configurations*. An active space is chosen which comprises the F_{2p}-like orbitals of the cluster and the projectile orbital(s) into which an electron can be transferred. All the orbitals of the active space can be unoccupied, singly, or doubly occupied. The occupation numbers define the different configurations of the system. One of these configurations has ionic character (positively charged projectile and all valence-band-states doubly occupied) while all other configurations have covalent character (projectile neutralized and a hole in the valence band). The MCSCF method solves self-consistently both for the orbital wavefunctions and the expansion coefficients at the same time. In a state-averaged calculation both the ground state (which is dominated by either the ionic or one of the covalent configurations) and several excited states are calculated simultaneously. The MCSCF method thus accounts – at least on a qualitative level – for the interaction between different electronic configurations. However, quantitatively correct results can only be achieved if the energetic ordering of the levels for large projectile distance is also properly described. As explained above, the latter requires the inclusion of hole screening. This, in turn, requires the inclusion of a prohibitively large number of configurations. Therefore, the energetic ordering of the ionic and covalent states may still be incorrect on the MCSCF level, as is the case for the system of H^+ colliding with a LiF surface (see Figure 10.3(a) below).

The description of screening effects can at least partially be achieved by a multi-reference configuration interaction (MR-CI) method. The many-electron wavefunction is expanded in terms of a number of excitations of *reference configurations* (customarily the configurations from the preceding MCSCF run). The expansion coefficients yielding the lowest energy are then determined while the orbital wavefunctions are kept constant. This allows the inclusion of many more configurations than in the MCSCF calculation. Through the virtual excitation of electrons into intermediate states, correlation of electrons within the active cluster is taken into account. However, only single and double excitations are included in the expansion – as the inclusion of higher excitations becomes computationally prohibitive for big systems. As a consequence, for larger clusters, the MR-CI method suffers strongly from the violation of size consistency, i.e. the correlation energy does not scale linearly with the number of atoms since only single and double excitations are taken into account. One may go beyond the MR-CI method by employing methods that account for size-consistency on an approximate level: the extended Davidson correction [15, 16], Møller–Plesset perturbation theory (MRPT), or the multi-reference averaged quadratic coupled cluster method (MR-AQCC) [17].

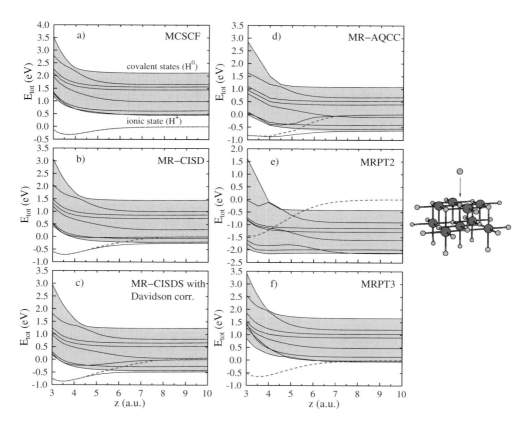

Figure 10.3: Potential energy curves for H^+ approaching an embedded $Li_{26}^+F_9^-$ cluster (vertical incidence, touch-down on Li site). Comparison of different levels of approximation: a) MCSCF, b) MR-CISD, c) MR-CISD with Davidson correction, d) MR-AQCC, e) MRPT2, f) MRPT3. The absolute energy scale is chosen such that the energy of the ionic state at large distance is 0. The dashed line indicates the *diabatic* energy curve corresponding to the ionic configuration. See also color figure on page 230.

We have tested the different quantum chemistry approaches for the calculation of potential energy surfaces and demonstrated that proper inclusion of correlation (i.e. screening of the holes) leads indeed to a proper energetic ordering of the levels of the system $H^+ \rightarrow$ LiF [3]. The results of our calculations are summarized in Figure 10.3 where we present the potential energy curves for H^+ in vertical incidence on top of the central Li^+ ion of an embedded $Li_{26}^+F_9^-$ cluster[5]. On the MCSCF level (Figure 10.3(a)), the lowest-lying state in energy corresponds at all distances to the ionic configuration. In the orbital energy picture this would mean that the capture level of H^+ lies above the valence band edge. As pointed out above, this

---

[5] The Li^+ ions and F^- ions are described with the same pseudo-basis [8] as in Section 10.2. Within the C_{2v} symmetry group, the F_{2p} atomic orbitals split into 8 valence orbitals of A_1-symmetry from which an electron can be transferred to the H^+ ion. These eight states represent the valence band of LiF shaded in grey.

wrong level ordering is due to the nearly complete neglect of screening effects on the MCSCF level. On the MR-CI level with single and double excitations (MR-CISD, Figure 10.3(b)), screening of the holes leads to a lowering of the binding energy of all covalent states (i.e. of all states where a hole is left behind in the surface) by 0.75 eV with respect to the ionic state. The shift due to the correlation energy leads to avoided crossings between the ionic entrance channel and some of the covalent states representing the exit channel. The dashed line indicates the diabatic energy curve of the ionic state which crosses several of the covalent curves. Since in large clusters the correlation energy is often underestimated, we also apply the Davidson correction [15, 16] to approximately correct for size consistency. The Davidson correction affects the covalent states more than the ionic state and leads to an additional downward shift in energy of the covalent states at large distance by 0.25 eV (Figure 10.3(c)). The ionic state is now clearly embedded into the "band" of covalent states. The energetic difference between the asymptotic ionic and lowest covalent level is 0.5 eV compared to the experimental value of 1.3 eV. A calculation on the MR-AQCC level (Figure 10.3 d) yields an even stronger asymptotic lowering of the covalent states. The resulting asymptotic energy difference between the lowest covalent and the ionic level is 0.63 eV and confirms the expectation that methods containing size consistency corrections such as AQCC should yield converged potential energy curves for charge exchange, provided that a calculation with larger cluster size and basis set becomes numerically feasible with further advances in computing power. For completeness, we present in Figure 10.3(e) and (f) calculations of the potential energy curves with Møller–Plesset perturbation theory to second order (MRPT2) and to third order (MRPT3). While MRPT2 leads to a considerable downward shift of the covalent levels by 1.8 eV, MRPT3 cancels this shift to a large extent and leads to a result similar to that of the MRCI-SD approximation. The large difference between MRPT2 and MRPT3 indicates that the perturbation series only slowly converges and higher order corrections should be taken into account. We presume that higher orders will lead again to a downward shift of the covalent levels and will eventually converge towards the result obtained by other methods such as the MR-AQCC method.

The screening effect is enhanced when larger active clusters are used but converges only slowly with cluster size since screening due to polarization of the environment is a long-range effect. It would therefore be desirable to combine the accurate, but computationally very demanding description of a small active cluster in the region around the point of projectile impact with a somewhat "cheaper" description of the larger environment which mainly contributes only through its polarizability. This leads us to the question which will be treated in the next section: if we describe the environment of the active cluster by static point charges and/or by a polarizable environment, how large must the active cluster itself be in order to properly describe the interaction of the projectile with the band structure.

10.4 Convergence of Potential Energy Curves as a Function of Cluster Size

Since the complex methods that properly describe screening are still prohibitively expensive for larger clusters, we have performed a convergence study of the embedded cluster method as a function of cluster size on the MCSCF level. This allows us to include a large number of reference configurations in order to explore the continuum limit of the valence states. Since the MCSCF method suffers – in principle – from a wrong level ordering for our sample system $H^+ \rightarrow LiF$, we can artificially enforce the correct level ordering by choosing a very small basis for the F^- ions[6]. The additional benefit of this small basis is that we can include large active clusters in our study. The embedded clusters of our study are shown in Figure 10.4. They range from a cluster containing only one active F^- up to a cluster with 13 active F^- in the topmost atomic layer. The active clusters are surrounded by an array of point charges such that the total system (active cluster and point charges together) consists of 196 ($7 \times 7 \times 4$) force centers. This renders the system neutral and reproduces the Madelung potential for an electron at the center site in the surface with sufficient accuracy.

As a first test we calculate the ionization potentials of different embedded clusters in the absence of the H^+ projectile through the energy difference ΔE between the total energy of the neutral systems and of the ionized systems[7]:

$$
\begin{aligned}
Li_5^+F^-: & \quad \Delta E = 9.61 \text{ eV}\\
Li_{17}^+F_5^-: & \quad \Delta E = 10.64 \text{ eV}\\
Li_{25}^+F_9^-: & \quad \Delta E = 10.65 \text{ eV}\\
Li_{37}^+F_{13}^-: & \quad \Delta E = 10.82 \text{ eV}
\end{aligned}
$$

In all cases, the ionization potential remains smaller than the ionization potential of hydrogen (13.6 eV). This corresponds to the correct level ordering in the presence of the projectile, i.e., the ionic state is higher in energy than the lowest covalent state. This correct ordering is the prerequisite for performing a convergence study with respect to cluster size on the MCSCF level.[8]

Figure 10.4 presents potential energy curves for the ionic and the covalent states of an H^+ ion impinging on clusters containing an increasing number of active F^- ions in the topmost surface layers. All F^- ions are fully coordinated by active Li^+ ions in order to prevent artificial distortion of the electron density at the border between the active cluster and the surrounding point charges. We present curves for the projectile in vertical incidence on top of a F^- ion in the surface layer. For this geometry, the complete system comprising the embedded cluster and the projectile is described by the C_{4v} symmetry group. The ionic state (neutral surface plus bare H^+) corresponds to a closed shell configuration and possesses therefore A_1 symmetry.

---

[6] The basis is taken from Ref. [18]. For F a contraction (10S,7P) $\rightarrow$ [2S,1P] is used and for Li a contraction (10s) $\rightarrow$ [2s]. The basis is so small that the electron affinity of fluorine calculated within this basis carries the wrong sign. This leads to an artificial upwards shift of the valence band by about 3 eV. For the present model calculation this shift is desired because it compensates for the absence of strong correlation effects in the MCSCF calculations.

[7] The positive system is calculated on the MCSCF level with a state average over all states that correspond to a hole in one of the F_{2p}-like orbitals. Correspondingly, the active space comprises all F_{2p}-like orbitals. The neutral systems are calculated on the SCF level (1 configuration with all F_{2p}-like orbitals doubly occupied). The ionization potential is calculated as the difference between the total energy of the lowest state of the ionized system and the total energy of the neutral cluster.

[8] However, we emphasize that the level ordering is only correct due to the artificially small basis chosen. Choosing a realistic basis will lead to much lower orbital energies of the F_{2p}-like orbitals corresponding to higher ionization potentials reaching up to 15 eV. This is because large basis sets including diffuse and polarization functions lead to a better accommodation of the electrons in the anionic state of the fluorines. The proper level ordering using a correct basis can only be restored by including a more sophisticated level of hole screening.

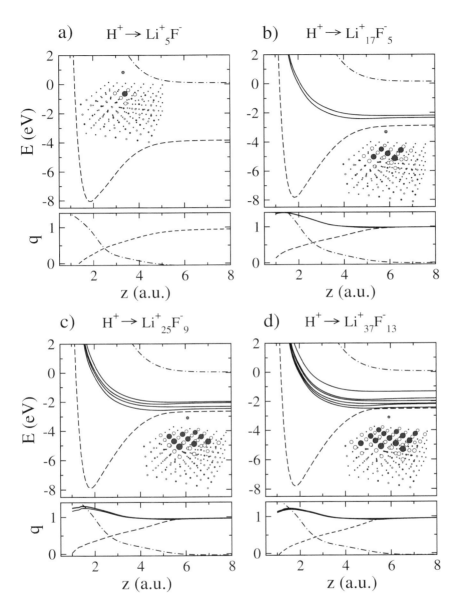

Figure 10.4: Potential energy curves for a H^+ ion interacting with embedded clusters of LiF of increasing size: a) $Li_5^+ F^-$, b) $Li_{17}^+ F_5^-$, c) $Li_{25}^+ F_9^-$, d) $Li_{37}^+ F_{13}^-$. Only curves of A_1 symmetry (within the C_{4v} symmetry group) are displayed. The insets display the clusters embedded into a lattice of point charges (black: F^- ions, white: Li^+ ions). In the lower panels we show the distance-dependent electronic charge q of the projectile. Dashed-dotted line: state with ionic character at large distance; dashed line: covalent state strongly interacting with the ionic state; solid lines: residual covalent states of A_1 symmetry. See also color figure on page 231.

Due to the Wigner–von Neumann non-crossing rule it can only interact with covalent states of the same symmetry. Therefore we show in Figure 10.4 only potential curves of A_1 symmetry.

In the smallest system containing only one active F^-, there are only two states of A_1 symmetry: one ionic and the other covalent (Figure 10.4(a)). At large projectile distance z, the two curves run in parallel with a distance in energy of 4 eV. At $z = 5$ a.u. the two curves start to separate. In order to determine the character of the avoided crossing, we show the (approximate) electronic charge localized at the hydrogen projectile for both states in the panel below. This charge can be easily calculated from the total electronic dipole of the system. A value of $q = 1$ at large distances signifies one electron located at the projectile and therefore characterizes the covalent state. Likewise, a value of 0 characterizes the ionic state. The lower state in the potential energy diagram (dashed line) has covalent character at large distance and the upper state (dash-dotted line) is ionic. This asymptotic energetic ordering is consistent with the fact that the calculated ionization energy of the embedded cluster (9.61 eV, see above) is lower than the ionization potential of hydrogen. At small distances, the two states exchange their character, as can be seen from the corresponding curve crossing in the charge diagram (lower panel). The lower state has now taken on ionic character which gives rise to the $1/z$-like slope before the curve reaches a minimum at 1.8 a.u. where the nuclear repulsion starts to dominate the interaction potential. Another way to verify that the two curves do indeed perform an avoided crossing is the analysis of the expansion coefficients of the MCSCF wavefunction. The ionic state at large distance corresponds to a configuration where the 1s orbital of hydrogen is unoccupied and the $2p_z$ orbital (with the z-axis perpendicular to the surface) of the fluorine is doubly occupied. In the covalent state, both orbitals are singly occupied. At small distances, the two atomic-like orbitals start to hybridize accompanied by a configurational mixing. At a distance of about 4 a.u. the two configurations contribute about 50% to each states which is a clear indication of an avoided crossing.

The addition of the four nearest neighbor fluorines to the active cluster in Figure 10.4(b) adds two additional states of A_1 symmetry. These states (solid lines) interact only weakly with the other two states. The corresponding potential energy curves are mostly flat (until the repulsive regime at small distances is reached) and have covalent character for all projectile distances as can be seen in the charge plot. The charge transfer proceeds between the two states marked by dashed and dashed-dotted lines. An analysis of the MCSCF wavefunctions reveals the underlying reason: the half-occupied molecular orbital of the strongly interacting covalent state is mostly localized at the central fluorine while in the other two states the half-occupied orbitals have a larger weight at the surrounding F^- ions.

Adding more fluorines to the active cluster (Figures 10.4(c) and (d)) does not change the emerging scenario that charge transfer is dominated by predominantly two channels. The additional levels of A_1 symmetry are almost independent of the distance, have delocalized wavefunctions, and remain covalent in character. This observation clearly indicates the suitability of the embedded cluster approach to describe the charge-transfer between a projectile ion and an extended surface: even though the capture state of the projectile can – in principle – interact with a continuum of states, in practice it only interacts with one state. For other scattering geometries, where the projectile is not incident on top of a fluorine, there may be several states interacting, but still only a small number of localized states is expected to contribute. The slope of the two states that represent the charge transfer channels at small distance (dashed and dashed-dotted lines) appears to be to converged as a function of cluster

a) z = 2 a.u.   b) z = 5 a.u.   c) z = 8 a.u.

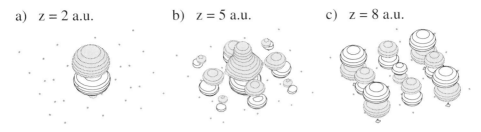

Figure 10.5: Orbital occupied by the hole in the main configuration contributing to the covalent state which interacts with the ionic state in Figure 10.4 at projectile distance a) $z = 2$ a.u. b) $z = 5$ a.u. c) $z = 8$ a.u. The points in the plots indicate the position of the nuclei (both F and Li) of the active cluster. White and grey indicate positive and negative values of the wavefunction.

size. Also the crossing point of the charge of these two states has converged with cluster-size. Figure 10.5 illustrates the localization of the interacting state for the system $H^+ \rightarrow Li_{25}^+F_9^-$, by displaying the wavefunction of the half-occupied molecular orbital that gives the dominant contribution to the interacting covalent state. It corresponds to the wavefunction of the hole left behind in the surface after transfer of an electron to the projectile. At large distances ($z = 8$ a.u.) the hole is almost evenly distributed over the 2p orbitals of all fluorines contained in the active cluster. At $z = 5$ a.u., the orbital is mostly localized at the central fluorine and shows a small admixture from the 1s orbital of the hydrogen projectile. At small distances ($z = 2$ a.u.), the hole is completely localized in a hybrid orbital comprising the $2p_z$ orbital of the central fluorine and the 1s hydrogen orbital. At this distance, the covalent configuration only contributes to the highest state (dashed-dotted line) in Figure 10.4(c) while at larger distances it contributes also to the lowest energy curve (dashed line). The analysis of the MC-SCF wavefunctions underlines the scenario that out of the many covalent states with a hole delocalized in the surface, one state localizes and represents the main charge transfer channel with the ionic state.

10.5 Conclusions

The accurate ab-initio treatment of charge-transfer in ion–surface collisions still poses a considerable computational challenge. Using the example of hydrogen ions impinging on a LiF surface, we have investigated in this chapter the feasibility of an approach where the (infinite) surface is represented by a finite embedded cluster only. With increasing cluster size, the discrete density of valence states of (bulk) embedded clusters converges towards the continuum DOS of LiF. The valence band of LiF is thus well represented by embedded clusters. We present a convergence study of potential energy curves for an H^+ ion interacting with clusters of increasing size. The projectile level can interact – in principle – with a continuum of valence states. An accurate description would then require embedded clusters of infinite size. In practice, however, our model calculations demonstrate that one, or at most a few, states localize in the region of impact as the projectile approaches the surface. The potential energy curves corresponding to these states clearly converge as a function of cluster size and display only weak interaction with the delocalized states. We have thus demonstrated that the embedded cluster approach is, indeed, feasible for the calculation of charge exchange in ion–surface collision. In practical calculations, the proper inclusion of correlation energy is important. Correlation effects lead to hole-screening, i.e., the polarization of the environment of a hole in the surface after transfer of an electron to the projectile. Within the region of the active cluster, correlation can be described to a good degree of approximation by size-consistent methods from quantum chemistry such as a multi-reference CI (including the Davidson correction) or coupled-cluster methods. A complete solution of the problem will, however, require the inclusion of polarization effects in the surrounding medium, at least on a phenomenological level. For the future, we plan the calculation of non-adiabatic coupling-matrix elements in order to solve the time-dependent Schrödinger equation and obtain cross sections for the neutralization of particles in ion–surface collisions.

Acknowledgments

This work was supported by the European Community Research and Training Network COMELCAN (HPRN-CT-2000-00128), and by EU project HPRI-2001-50036, FWF-SFB016, and P14442-CHE. We acknowledge stimulating discussion with S. Pantelides about the feasibility of finite clusters for the representation of extended surfaces.

References

[1] F. Aumayr, J. Burgdörfer, P. Varga, and HP. Winter, Comments At. Mol. Phys. **34**, 201 (1999).

[2] P. Fulde, *Electron Correlations in Molecules and Solids*, Springer, New York, 1995.

[3] L. Wirtz, J. Burgdörfer, M. Dallos, T. Müller, and H. Lischka, Phys. Rev. A **68**, 032902 (2003).

[4] COLUMBUS, *An ab initio Electronic Structure Program*, Release 5.7 (2000), written by: H. Lischka, R. Shepard, I. Shavitt, F.B. Brown, R.M. Pitzer, R. Ahlrichs, H.-J. Böhm, A.H.H. Chang, D.C. Comeau, R. Gdanitz, H. Dachsel, M. Dallos, C. Erhard, M. Ernzerhof, G. Gawboy, P. Höchtl, S. Irle, G. Kedziora, T. Kovar, Th. Müller, V. Parasuk, M. Pepper, P. Scharf, H. Schiffer, M. Schindler, M. Schüler, E. Stahlberg, P.G. Szalay and J.-G. Zhao.

[5] G. Hayderer, M. Schmid, P. Varga, HP. Winter, F. Aumayr, L. Wirtz, C. Lemell, J. Burgdörfer, L. Hägg, and C. O. Reinhold, Phys. Rev. Lett. **83**, 3948 (1999).

[6] L. Wirtz, G. Hayderer, C. Lemell, J. Burgdörfer, L. Hägg, C. O. Reinhold, P. Varga, HP. Winter, and F. Aumayr, Surf. Sci. **451**, 197 (2000)

[7] TURBOMOLE, *Program Package for ab initio Electronic Structure Calculations*, Version 4 (1997), written by R. Ahlrichs et al.

[8] W. J. Stevens, H. Basch, and M. Krauss, J. Chem. Phys. **81**, 6026 (1984).

[9] F. J. Himpsel, L. J. Terminello, D. A. Lapiano-Smith, E. A. Eklund and J. J. Barton, Phys. Rev. Lett. **68**, 3611 (1992).

[10] E. L. Shirley, L. J. Terminello, J. E. Klepeis, and F. J. Himpsel, Phys. Rev. B **53**, 10296 (1996).

[11] X. Gonze, J.-M. Beuken, R. Caracas, F. Detraux, M. Fuchs, G.-M. Rignanese, L. Sindic, M. Verstraete, G. Zerah, F. Jollet, M. Torrent, A. Roy, M. Mikami, Ph. Ghosez, J.-Y. Raty, D.C. Allan, Comput. Mater. Sci. **25**, 478-492 (2002).

[12] Y. Wang, P. Nordlander, and N. H. Tolk, J. Chem. Phys. **89**, 4163 (1988).

[13] A. B. Kunz, Phys. Rev. B **26**, 2056 (1982).

[14] R. Shepard, *The Multiconfiguration Selfconsistent Field Method*, volume LXIX of *Advances in Chemical Physics, Ab Initio Methods in Quantum Chemistry II*, pp. 63-200, Wiley, 1987.

[15] S. R. Langhoff and E. R. Davidson, Int J. Quantum Chem. **8**, 61 (1974).

[16] P. J. Bruna, S. D. Peyerimhoff, and R. J. Buenker, Chem. Phys. Lett. **72**, 278 (1981).

[17] P. G. Szalay and R. J. Bartlett, Chem. Phys. Lett. **214**, 481 (1993); P. G. Szalay and R. J. Bartlett, J. Chem. Phys. **103**, 3600 (1998).

[18] E. A. García, P. G. Bolcatto, M. C. G. Passeggi, and E. C. Goldberg, Phys. Rev. B **59**, 13370 (1999).

11 Coincident Studies on Electronic Interaction Mechanisms during Scattering of Fast Atoms from a LiF(001) Surface

Helmut Winter

The electronic interaction mechanisms during grazing scattering of hyper-thermal hydrogen atoms from a LiF(001) surface are studied via projectile-electron coincidences. This type of translation energy spectroscopy applied to atom–surface scattering allows one to investigate in detail the relevant electronic processes. It turns out that the transient formation of negative ions during the scattering process is the precursor for electronic excitations of the solid as well as for electron emission to vacuum. The specific nature of electron transfer in terms of a local capture event from the anion sites of the ionic crystal plays a decisive role in the efficient formation of negative ions. Our studies unravel the complete reaction paths during the scattering process and provide a consistent interpretation for the, at first glance, surprisingly large total electron emission yields for impact of atoms/ions on the surface of ionic crystals.

11.1 Introduction

Insulator surfaces play an important role in the efficient conversion of energy of microscopic projectiles to electron emission. Total electron emission yields for bombardment of solid surfaces by electrons of some 100 eV show clearly higher yields for oxidized targets than their clean-metal counterpart [1]. Similar results were observed with fast atoms and ions [2,3]. Here we mention, in particular, relatively high total electron yields and low kinetic thresholds for ion/atom impact on surfaces of ionic crystals [4, 5]. As a recent example of this feature we show in Figure 11.1 total electron yields as a function of projectile velocity for grazing scattering of fast hydrogen atoms from a clean and flat LiF(001) (full circles) and an Al(111) surface (open circles). Details of the measurements will be outlined below. Despite clearly higher binding energies of electrons in LiF available for emission and charge transfer, one finds higher total electron emission yields for impact on LiF than for the metal target. Also a kinetic threshold occurs at smaller energies for the ionic crystal than for the metal.

For an estimate of electron emission from a solid target we consider the maximum in energy which can be transferred to an electron of kinetic energy E_e and mass m_e in a classical binary collision with an atomic projectile of mass M. From conservation of momentum and energy, one finds for an electronic binding energy with respect to vacuum E_b a threshold for

Correlation Spectroscopy of Surfaces, Thin Films, and Nanostructures. Edited by Jamal Berakdar, Jürgen Kirschner
Copyright © 2004 Wiley-VCH Verlag GmbH & Co. KGaA, Weinheim
ISBN: 3-527-40477-5

the projectile energy [6]

$$E_{\text{th}} = \frac{M}{2\,m_{\text{e}}} \left(E_{\text{e}} - \sqrt{E_{\text{e}}\,(E_{\text{e}} + E_{\text{b}})} + \frac{E_{\text{b}}}{2} \right) \tag{11.1}$$

This amounts for collisions of H atoms with an Al surface ($E_{\text{e}} = 10.6$ eV = Fermi energy, $E_{\text{b}} = W = 4.3$ eV = work function) to $E_{\text{th}} = 168$ eV ($v_{\text{th}} = 0.082$ a.u.; a.u. = atomic unit), whereas for LiF ($E_{\text{e}} \approx 4$ eV = width of F2p valence band, $E_{\text{b}} = 12$ eV) we find $E_{\text{th}} = 1836$ eV ($v_{\text{th}} = 0.271$ a.u.). The resulting E_{th} for the insulator is about one order of magnitude higher than for the metal surface which is in sheer contrast with the experimental findings.

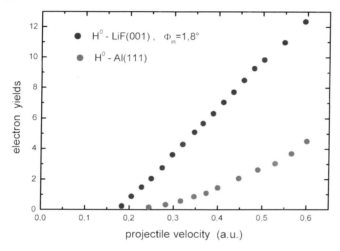

Figure 11.1: Total electron emission yields as a function of projectile velocity for scattering of hydrogen atoms from LiF(001) (upper curve) and Al(111) (lower curve) under a grazing angle of incidence $\Phi_{\text{in}} = 1.8°$. See also color figure on page 232.

In the following we will discuss recent developments in the study of electronic excitation and emission processes induced by impact of atomic projectiles on an insulator surface. We will show that the coincident detection of the energy loss of scattered projectiles with the number of emitted electrons provides specific information on the electronic interaction mechanisms. On the basis of these experiments a consistent interpretation of the high total electron emission yields can be derived in terms of a reaction scheme, where the transient formation of negative ions in local electron capture events from anion sites plays a key role.

11.2   Experimental Developments

In Figure 11.2 we give a sketch of the experimental setup which was used for the investigation discussed here. In an UHV chamber at a base pressure of some 10^{-11} mbar H atoms are scattered under a grazing angle of incidence Φ_{in} of typically $1°$ from a clean and flat LiF(001) target. The fast H atoms with energies ranging from some 100 eV to some 10 keV are produced via (near-resonant) charge exchange in a gas target operated with Kr atoms. Three sets of slits provide a collimation of the projectile beam in the sub-mrad regime and are, furthermore, used as components of two differential pumping stages. Before entering the gas target, the ions are deflected by a pair of electric field plates, biased with voltages of rise times in the ns-domain. This device is used as beam chopper for time-of-flight (TOF) measurements with an overall time resolution of some ns (see below).

The LiF(001) target was mechanically polished and prepared in UHV by a fair number of cycles of grazing sputtering with Ar^+ ions at temperatures of about 300 °C and subsequent heating to about 400 °C. Angular distributions of scattered projectiles are recorded by means of a channeltron detector, and a channelplate is used as start detector for the TOF-device.

Electrons emitted during projectile impact at the target are recorded by a surface barrier detector (SBD), kept on a potential of about 25 kV. A highly transparent grid (transmission about 98%) is biased to some 10 V for collection of electrons from the target region and to shield the high potential from the target surface. After electrons have passed the grid, they are accelerated and focused on the entrance of the SBD and generate a pulse with a height proportional to the energy of all incident electrons. Since the overall time resolution of this detector is poor (about μs), a pile up of pulses from electrons which stem from the same emission event takes place; then the pulse heights of the detector are proportional to the number of electrons emitted during the scattering of individual projectiles [7–10].

An interesting feature of the setup shown in Figure 11.2 is the combination of the TOF-device with the SBD, i.e. the coincident combination of translation energy spectroscopy with the number of emitted electrons. This technique allows one to relate the overall electronic excitations during the scattering event to the emission of a specific number of electrons.

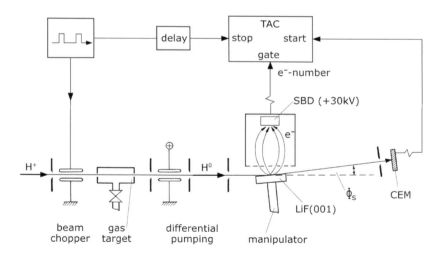

Figure 11.2: Sketch of experimental setup for coincident TOF-electron number studies.

11.2.1 Energy Loss Spectroscopy via Time-of-Flight

A main issue for our studies is the use of neutral atomic projectiles, i.e. H atoms here. Since ions may give rise to potential emission (PE) of electrons, possibly affected by kinematic effects, one should select neutral ground state atoms for studies on kinetic emission (KE) of electrons. A further reason for using neutral particles in the incidence and exit channel is related to the experimental condition of a biasing electric field for collection of electrons at the target surface. In the grazing incidence collision geometry, such a field would deflect charged projectile beams from the target surface in a manner difficult to control.

Figure 11.3 shows spectra for He^+ ions and He^0 atoms of 2 keV energy where the atoms are produced in the gas target operated with He via resonant charge transfer. The spectra are recorded with a standard TOF design with an overall time resolution of about 5 ns. This relates to a half width of the energy of about 5 eV. Note that the atoms lose typically 5 eV of energy in the neutralization process owing to transfer of energy to target atoms in the gas phase collisions. The spectrum shows, furthermore, that excitations to excited states of He atoms (excitation energies of metastable 2s levels amount to about 20 eV) cannot be resolved within the sensitivity level of detection. We will show below that a good energy resolution is essential for studies on electronic excitations of insulators during grazing ion/atom surface scattering.

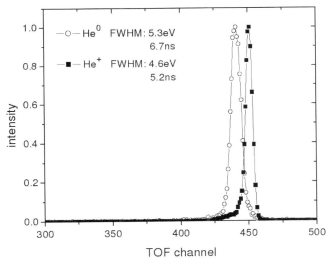

Figure 11.3: TOF spectra for 2 keV He^0 (open symbols) and He^+ (full symbols) projectiles. Atoms are produced in the gas target in the beam line operated with He atoms.

11.2.2 Electron Number Spectra

The number of emitted electrons for impact of an atom/ion on the target surface is recorded by means of a SBD biased with a voltage of about 25 kV with respect to ground. As an example we display in Figure 11.4 a spectrum of the SBD for scattering of 16 keV He atoms from an Al(111) surface under $\Phi_{\text{in}} = 1.6°$. In the spectrum (logarithmic scale) peaks due to a specific number of emitted electrons are clearly identified (upper trace), however, at low pulse heights substantial noise is present as demonstrated by a spectrum (lower trace) recorded for no bombardment of the target. This method of electron detection allows one to investigate the electron emission statistics and to obtain precise total electron emission yields γ from measured probabilities W_n for a specific number n of emitted electrons [10] via

$$\gamma = \sum_{0}^{\infty} n\, W_n \left/ \sum_{0}^{\infty} W_n \right. \tag{11.2}$$

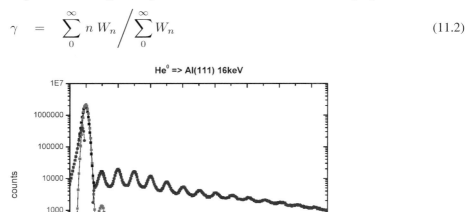

Figure 11.4: Non-coincident pulse height spectra of SBD referred to electron number for scattering of 16 keV He^0 from Al(111) under $\Phi_{\text{in}} = 1.9°$ (upper trace). Lower trace: background signal. See also color figure on page 232.

For low γ W_o will dominate the electron number spectrum. Since with a free running SBD an information on the emission of no electrons cannot be obtained (low peak heights stem from detector noise), the determination of total yields in this regime of operating the detector can only be obtained indirectly. A coincident detection with scattered projectiles, however, can solve this problem. Figure 11.5 shows the SBD spectrum from Figure 11.4 after subtraction of a background signal for low n which causes problems within the noise level of the detector (note that difference does not cancel to zero at low peak heights).

The lower trace in Figure 11.5 represents data for recording the SBD pulse in coincidence with scattered projectiles. In order to directly relate events of scattered projectiles at the channelplate with emitted electrons, one has to gate the SBD signal in a specific time window, equivalent to the flight time of projectiles from target to channelplate, i.e. one has to apply a time-of-flight (TOF) method. The "trigger" for the TOF studies has to be independent of electron detection, since otherwise events related to the emission of no electrons will be missed. As a consequence, an independent TOF branch with beam copper is installed which is operated as a gate control for the SBD detector. Then each event recorded in the TOF setup will be related to SBD detector peak heights, where heights equivalent to the noise level of the detector can be ascribed to events without emission of an electron. The coincident SBD pulse height spectrum (cf. Figure 11.5) is free from noise and shows a peak around zero (note small offset owing to small bias voltage at detector output). From such spectra reliable total electron yields can be derived [11].

For grazing scattering this type of electron detection has a further crucial advantage over conventional methods (current measurements). This is based on the feature that yields are derived from single impact events which is widely independent of the beam-target geometry. Yields derived from current measurements are affected by the target geometry which can introduce uncertainties owing to the small projection of the effective target surface onto the plane normal to the incident projectile beam. In particular, at small grazing angles of incidence this feature is important and can easily introduce considerable uncertainties for γ.

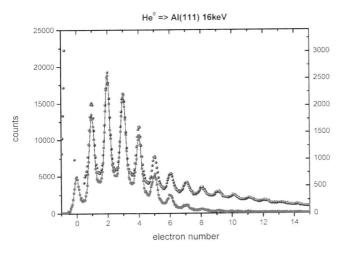

Figure 11.5: Coincident (lower trace) and non-coincident (upper trace) electron number spectra for scattering of 16 keV He^0 from Al(111) under $\Phi_{in} = 1.9°$. See also color figure on page 233.

11.3   Coincident TOF and Electron Number Spectra

With the setup shown in Figure 11.2 we recorded TOF spectra in coincidence with the electron channel, where a specific discriminator interval for the SBD pulses was used to separate contributions related to the emission of a selected number of electrons. As an example, we show in Figure 11.6 TOF spectra for the scattering of 1 keV H atoms scattered from a LiF(001) surface under a glancing angle of incidence $\Phi_{in} = 1.6°$. The TOF spectra are characterized by discrete peaks which reflect the discrete projectile energy loss due to the effect of the wide band gap of LiF (about 14 eV) on electronic excitations. For the spectrum related to the emission of no electron (noise level of SBD) one observes a prominent peak ascribed to elastic scattering of projectiles with surface atoms. Recently we have studied this so called "nuclear energy loss" under surface channeling conditions [12] with Ne atoms and estimate a contribution of less than 1 eV for the present case.

An at first glance surprising feature is the second peak in this spectrum, since it cannot be related to the emission of an electron. In similar studies, Roncin et al. [13] have identified this peak as due to an internal excitation of the target surface, a surface exciton, i.e. a local electron hole pair formation at anion lattice sites (F^- here). The third peak in this spectrum stems from the excitation of a further surface exciton, etc.

The TOF spectrum related to the emission of one electron shows the expected offset with respect to its first peak, since for the emission of an electron from the valence band of LiF at least the binding energy of 12 eV has to be transferred from the projectile to an electron. We find a mean energy of 14 eV which is consistent with a width of the valence band of a few eV and a finite energy of the emitted electron. Note the clear shift in energy between the two peaks related to emission of an electron and excitation of a surface exciton (12 eV). The further peaks observed in the TOF spectrum for one electron are ascribed to the production of additional excitons. The spectrum for the emission of two electrons shows a further shift towards a higher energy loss, since two electrons have to be lifted from the valence band of LiF to vacuum.

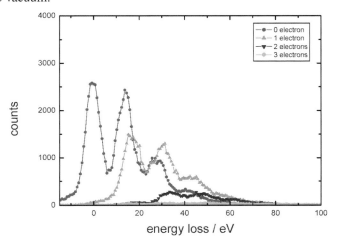

Figure 11.6: TOF spectra (converted to energy loss) coincident with number of emitted electrons for scattering of 1 keV hydrogen atoms from LiF(001) under $\Phi_{in} = 1.8°$. For details see text. See also color figure on page 234.

This version of translation energy spectroscopy [14] allows one to investigate the relevant electronic excitation phenomena during projectile impact. The coincident detection of electrons provides, furthermore, information on the fraction of inelastic processes that lead to the emission of electrons. In order to analyze the data in this respect, it is favorable to handle the spectra irrespective of the settings of discriminator levels (cf. Figure 11.6). An experimental solution in this respect is the recording of the full TOF *and* SBD spectra in terms of a 2D array. In Figure 11.7 we show a resulting 2D plot of spectra as discussed for selected SBD pulse heights in Figure 11.6.

These 2D spectra provide a profound basis for the identification of the excitation mechanisms and evaluation of data. Peaks in the left column (0 electrons) can be ascribed to elastically scattered projectiles, production of one exciton, two excitons; second column (1 electron): emission of one electron, emission of one electron plus production of one exciton, etc.; third column (2 electrons): emission of two electrons, emission of two electrons plus production of one exciton, etc. Note that, in particular, for the signals related to the emission of more than one electron weak peaks appear in the spectra for energy losses necessary to emit one electron. This "crosstalk" might be caused by additional spurious electrons produced within the vacuum chamber and by a small portion of pulses of the chopped projectile beam which contains more than one fast atom. It turns out that this problem limits the application of this new powerful method in studies on threshold effects for electron emission. Nevertheless, we will show below that this coincident detection provides reliable total electron emission yields down to the sub-permille domain. For specific cases yields as low as some 10^{-6} have been detected by this method [15].

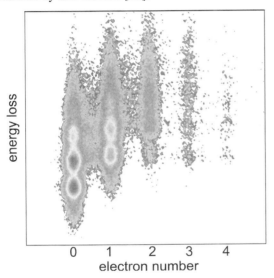

Figure 11.7: 2D plot of coincident TOF (vertical axis, "energy loss") and SBD spectra (horizontal axis, "electron number") for scattering of 1 keV hydrogen atoms from LiF(001) under $\Phi_{\text{in}} = 1.8$ deg.

In Figure 11.8 we show the result of an analysis of data from Figure 11.7 in terms of relative peaks heights for emission of 0, 1 and 2 electrons and production of a number of excitons. The full bars represent the experimental data, the open bars result from a description of data on the basis of a model outlined in Section 11.4.

11.3.1   Studies on Near-Threshold Behavior

In the previous section we have already mentioned that the coincident detection of TOF and electron number spectra provides, in addition to a free-running SBD, direct information on

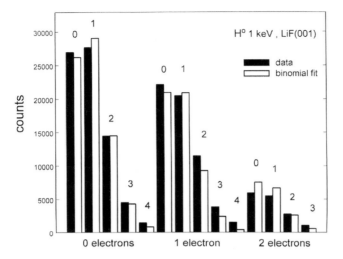

Figure 11.8: Bar graph of intensities for spectra shown in Figure 11.7 for emission of a specific number of electrons and production of excitons (number at bar). Full bars: experiment, open bars: analysis of data in terms of binomial statistics with parameters derived from interaction model.

the probability W_o for the emission of no electrons. This makes this technique an ideal tool for studies of low total electron emission yields γ and, in particular, for investigations near respective thresholds.

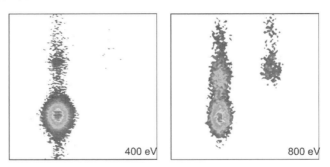

Figure 11.9: 2D plot of coincident TOF and SBD spectra for scattering of 400 eV (left) and 800 eV (right) hydrogen atoms from LiF(001) under $\Phi_{\text{in}} = 1.8°$. See also color figure on page 234.

In Figure 11.9 we show 2D spectra for H atoms of energy 400 eV (left panel) and 800 eV (right panel) scattered from LiF(001). A qualitative comparison of the spectra displayed in Figures 11.7 and 11.9 indicates clearly that the probabilities for electronic excitations are strongly reduced for decreasing projectile energy. At 400 eV the spectrum is dominated by the peak stemming from elastically scattered projectiles, only a small portion of excitons are produced and the signal from emission of electrons is extremely weak and close to the noise level of detection.

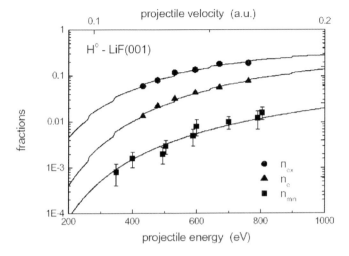

Figure 11.10: Exciton (full circles), electron (full triangles), and H^- fractions (full squares) as a function of projectiles energy for scattering of hydrogen atoms from LiF(001) with constant normal energy $E_z = 0.6$ eV.

The probabilities for exciton production (full circles) and electron emission (full triangles) derived from 2D spectra are displayed in Figure 11.10 as a function of projectile velocity with constant normal energy $E_z = 0.6$ eV. We plot also the fractions of negative H^- ions in the scattered beam as measured with an additional channeltron detector positioned about 50 cm behind the target and electric field plates in order to disperse the beam with respect to charge.

The probabilities are generally small (< 0.1) and decay with decreasing projectile energy (velocity). The solid curves represent results from an analysis of the data in terms of a model for the electronic interaction mechanisms described in the following section.

11.4   Model for Electronic Excitation and Capture Processes during Scattering of Atoms from Insulator Surfaces

The flat valence band of an alkali metal halide crystal is formed by electrons from the ionic bond which are fairly located at halogen lattice sites. Binding energies and band gaps amount to typically 10 eV. An important development for understanding the processes studied here was the experimental finding of fairly large negative ion fractions after the scattering of reactive atoms from alkali metal halide surfaces under glancing angles of incidence [16,17]. These results were explained by a local capture process of electrons from anion sites ("active sites" concerning charge transfer), where the large energy defect between initial and final states in a binary type of collision is reduced by the effect of the crystal lattice [18]. Electrons interacting with active sites are affected by the Madelung potential so that the low binding energies of negative ions (typically eV) are substantially increased. This interaction scenario is sketched in Figure 11.11.

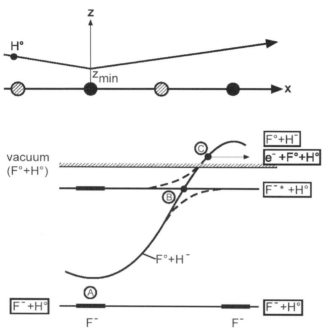

Figure 11.11: Sketch of the projectile trajectory (upper part) and the interaction model (lower part) comprising potential energy curves for scattering of hydrogen atoms from LiF(001) under grazing angle of incidence.

At active sites the formation of a negative ion may take place mediated by the confluence of levels for the two collision partners (labeled "A" in the sketch of the potential energy curves in Figure 11.11). The probability of a transition in this binary collision can be estimated from Demkov theory [19]

$$P_{\text{bin}} = \frac{1}{2} \sec h^2 \left(\frac{\pi \, \alpha \, \Delta E}{2 \, v} \right) \tag{11.3}$$

with $1/\alpha = (E_1^{1/2} + E_2^{1/2})/2^{1/2}$, E_1, E_2 being the electron binding energies of the collision partners and ΔE the effective energy difference between potential curves (energy defect).

After formation of a negative ion and escape from an active site the diabatic potential curve for $(H^- + F^0)$ crosses the exciton level $(H^0 + F^{-*})$. For this case the transition probability can be derived from Landau–Zener theory [20]

$$P_{\text{LZ}} = \exp \left(\frac{\pi \, \Delta \varepsilon_x^2}{2 \, (\text{d}V/\text{d}R)_x \, v} \right) \tag{11.4}$$

with $\Delta\varepsilon_x$ being the energy gap of the adiabatic and $(dV/dR)_x$ the difference in the slopes at the diabatic crossing point of the diabatic potential energy curves. Negative ions that have survived the crossing with probability P_{LZ} may detach thereafter with probability P_{det} before the next active site is reached. A complete reaction scheme is presented in Figure 11.12.

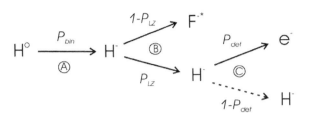

Figure 11.12: Reaction scheme for electronic interactions of hydrogen atoms with LiF. See also Figure 11.11.

Based on the reaction scheme given in Figure 11.12, the modified (n'_o and n'_{min}) occupation of a neutral atom n_o and negative ion n_{min} (normalization $n_o + n_{min} = 1$) after one interaction sequence is

$$n'_o = n_o\left(1 - P_{bin}\right) + n_o P_{bin}\left[(1 - P_{LZ}) + P_{LZ} P_{det}\right] + n_{min} P_{det} \tag{11.5}$$

$$n'_{min} = n_o P_{bin} P_{LZ}\left(1 - P_{det}\right) + n_{min}(1 - P_{det}) \tag{11.6}$$

For this sequence, the probability for emission of an electron n_e and production of a surface exciton n_{ex} is given by

$$n_e = n_o P_{bin} P_{LZ} P_{det} + n_{min} P_{det} \tag{11.7}$$

$$n_{ex} = n_o P_{bin}\left(1 - P_{LZ}\right) \tag{11.8}$$

The reaction products observed in our experiments are analyzed in terms of Eqs. (11.4)–(11.7). The final occupations and yields for electron emission and production of excitons are obtained from the iteration over N_{coll} effective collisions (number of active sites of the crystal being passed by the projectiles). Since complete potential energy curves for the present system and details on the detachment mechanisms are still a matter of debate, we used the relevant quantities for the description of probabilities as free parameters, i.e. $\alpha\Delta E$, $\beta = \pi\Delta\varepsilon_x^2/2(dV/dR)_x$, and P_{det}.

For a glancing angle of incidence Φ_{in} the motion of projectiles parallel to the surface (x-axis) proceeds with energy $E_x = E\cos^2(\Phi_{in}) \approx E$, the normal motion with $E_z = E\sin^2(\Phi_{in})$. The trajectory length is $\Delta s = \Delta z/\sin(\Phi_{in}) \approx v_z\Delta t/\Phi_{in}(\Delta z = v_z\Delta t$ is interval along z that relates to Δs) and for a constant normal velocity component $v_z = v\sin(\Phi_{in})$ trajectory lengths scale as $1/\Phi_{in} \sim v/v_z \sim (E/E_z)^{1/2}$. This feature is important for a variation of the projectile velocity v (energy E) where the normal velocity component v_z (energy E_z) is kept constant via an appropriate tuning of the angle of incidence Φ_{in}.

The solid curves in Figure 11.10 represent best (correlated) fits to the data using Eqs. (11.4)–(11.7), where the functional form of P_{bin} and P_{LZ} is taken from Eqs. (11.3) and (11.4). P_{det} is considered as constant for these projectile energies. The number of effective collisions N_{coll} is adjusted to the trajectory length. Variation of N_{coll} yields lowest χ^2 for about 10 collisions, furthermore we find $\alpha\ \Delta E = 0.22$ a.u., $\beta = 0.21$ a.u., and $P_{det} = 0.50$.

With these parameters, the data in Figure 11.9 are fairly well reproduced. This holds also for the description of the statistical analysis of data presented in Figure 11.8.

Further information on the electron detachment can be derived from the fractions of negative ions in the scattered beam. In Figure 11.13 we show negative ion fractions for the present case as a function of projectile velocity ($v = 0.2$ a.u. corresponds to a projectile energy of 1 keV). A kinematic resonance structure is observed with a maximum for negative ions of about 7%.

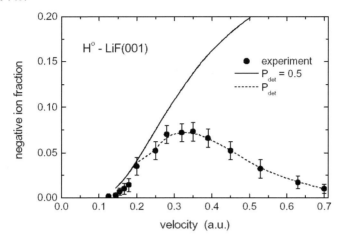

Figure 11.13: Negative ion fractions for grazing scattering of hydrogen atoms from LiF. Experiment: full circles; model with $P_{\text{det}} = 0.5$ (solid curve); model with P_{det} adjusted in order to match experimental data (dashed curve), cf. Figure 11.14.

The solid curve in the figure shows the negative ion fractions calculated from the model with the parameters given above. These calculations clearly give too large ion fractions for velocities $v > 0.3$ a.u. (projectile energies > 2 keV). Such an increase with projectile velocity is observed for the formation of negative ions with higher affinity energies than hydrogen during scattering from LiF (O^-, Cl^-, F^-, etc. [17]). We interpret the data by enhanced detachment probabilities. The experimental H^- fractions can be reproduced by an adjustment of P_{det} as a function of projectile velocity as shown by the solid curve in Figure 11.14.

The increase in P_{det} with velocity is the origin for the peaked structure of the experimental negative ion fractions shown in Figure 11.13. Whereas P_{bin} and P_{LZ} as given by Eqs. (11.3) and (11.4) approach 1 in the high velocity limit, i.e. complete formation of negative ions, the increasing P_{det} will reduce the fractions of surviving ions. Large P_{det} for negative hydrogen ions explains the relatively small negative ion fractions in the scattered beam in comparison to other reactive atoms, where larger affinity energies lead to smaller P_{det} and higher yields for negative ions (e.g., 80% for F^- formation for scattering from LiF [17]).

Borisov and Gauyacq [21] have investigated electron detachment for H^- ions in front of a LiF surface with a wave-packet-propagation method. In Figure 11.14 we show electron loss rates as a function of projectile velocity at $z = 2.5$ a.u. and different azimuthal directions. Despite the fact that a quantitative comparison is beyond the scope here, since the interaction

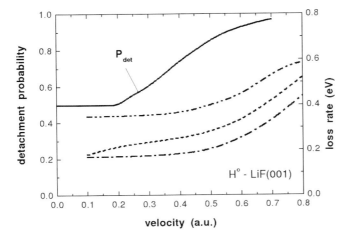

Figure 11.14: Detachment probability P_{det} as a function of projectile velocity adjusted in order to match the data in Figure 11.13. Dashed curve: theoretical loss rate for H^--LiF at distance $z = 2.5$ a.u. along $<100>$ above rows; dashed-dotted curve: between rows; dot-dot-dashed curve: along $<110>$ above F^- rows.

scenario has to be followed over complete trajectories, the theoretical loss rates show about the same kinematic dependence as the detachment probabilities derived from the analysis of the experiments. The main mechanism for electron detachment is attributed to binary interactions between projectiles and adjacent F^- sites of the crystal lattice.

11.5   Summary and Conclusions

In this chapter we have demonstrated that the coincident combination of translation energy spectroscopy with the number of emitted electrons allows one to obtain detailed information on the electronic excitation and emission mechanisms for (grazing) impact of atomic projectiles on solid surfaces. The application of this method to the scattering of hydrogen atoms from a LiF surface (wide-band-gap insulator) has led to a basic understanding of the interaction processes for electron emission, internal excitations of the target (surface excitons), and formation of negative ions. A basic feature is the (transient) formation of negative ions in local capture events of electrons from anion sites as the dominant precursor for the electronic excitations. Furthermore, the wide band gap will affect transport of electrons within the bulk and at the surface of ionic crystals.

This interaction mechanism is clearly different from the transfer of projectile energy in direct atom-electron encounters which is the dominant mechanism for electron emission from metal surfaces near threshold [22]. Coming back to our introductory statements, the fairly efficient mechanisms for electronic excitations in atom collisions with insulators provide a natural explanation for higher electron yields observed for insulators in comparison to metal targets.

Acknowledgments

The contributions to the work presented here from S. Lederer (HU Berlin) and the fruitful collaboration with Professors F. Aumayr and HP. Winter (TU Vienna) is gratefully acknowledged. The studies were supported by the Deutsche Forschungsgemeinschaft (DFG) under contract Wi1336/5 and Austrian FWF (project 14337-PHY).

References

[1] M. Ardenne, *Tabellen zur angewandten Physik* (Deutscher Verlag Wiss., Berlin, 1973), Band I, p.97.

[2] D. Hasselkamp, *Particle Induced Electron Emission II* (Springer, Heidelberg 1992), Springer Tracts Mod. Phys., Vol. 123, p. 1.

[3] J. Schou, Scanning Microsc. **2**, 607 (1988).

[4] E.W. Blauth, W.M. Draeger, J. Kirschner, N. Müller, and E. Taglauer, J. Vac. Sci. Technol. **8**, 384 (1991).

[5] P.A. Zeijlmans van Emmichoven, A. Niehaus, P. Stracke, F. Wiegershaus, S. Krischok, V. Kempter, A. Arnau, F.J. Garcia de Abajo, and M. Penalba, Phys. Rev. B **59**, 10950 (1999).

[6] R.A. Baragiola, E.V. Alonso, and A. Oliva Florio, Phys. Rev. B **19**, 121 (1979).

[7] K.H. Krebs, Ann.Phys. (Leipzig) **10**, 213 (1962).

[8] G. Lakits, F. Aumayr, and HP. Winter, Rev. Sci. Instrum. **60**, 3151 (1989).

[9] F. Aumyar, G. Lakits, and HP. Winter, Appl. Surf. Sci. 63, **17** (1991).

[10] G. Lakits, F. Aumayr, M. Heim, and HP. Winter, Phys. Rev. A **42**, 5780 (1990).

[11] A. Mertens, K. Maass, S. Lederer, H. Winter, H. Eder, J. Stöckl, HP. Winter, F. Aumayr, J. Viefhaus, and U. Becker, Nucl. Instrum. Methods B **182**, 23 (2001).

[12] A. Mertens and H. Winter, Phys. Rev. Lett. **85**, 2825 (2000).

[13] P. Roncin, J. Vilette, J.P. Atanas, and H. Khemliche, Phys. Rev. Lett. **83**, 864 (1999).

[14] H.B. Gilbody, Adv. At. Mol. Opt. Phys. **32**, 149 (1994).

[15] S. Lederer, K. Maass, D. Blauth, H. Winter, HP. Winter, and F. Aumayr, Phys. Rev. B **67**, 121405 (R) (2003)

[16] C. Auth, A.G. Borisov, and H. Winter, Phys. Rev. Lett. **75**, 2292 (1995).

[17] H. Winter, Progr. Surf. Sci. **63**, 177 (2000).

[18] A.G. Borisov, V. Sidis, and H. Winter, Phys. Rev. Lett. **77**, 1893 (1996).

[19] Y.N. Demkov, Sov. Phys. JETP **18**, 138 (1964).

[20] R.E. Johnson, *Atomic and Molecular Collisions* (Plenum press, New York, 1982), p. 116.

[21] A.G. Borisov and J.P. Gauyacq, Phys. Rev. B **62**, 4265 (2000).

[22] H. Winter and HP. Winter, Europhys. Lett. **62**, 739 (2003).

12 Many-body Effects in Auger–Photoelectron Coincidence Spectroscopy

Masahide Ohno

By Auger–photoelectron coincidence spectroscopy (APECS) we can manipulate the measurements of initial and final products of core-level electron photoionization. As a result, we can study the correlations between initial and final events. Moreover, by the coincidence measurement, we may be able to select the physical quantity of interest, such as the initial core-hole self-energy, so that it predominantly governs the coincidence spectrum. As an example, we discuss a many-body effect on a coincidence spectrum by shakeup/down excitations. We explain the recently observed line shift and narrowing of the coincidence L_3 photoelectron line of Cu metal.

12.1 Introduction

Photoelectron spectroscopy and Auger-electron spectroscopy (AES) are powerful means to study chemical compositions and electronic structures of surfaces, atoms and molecules adsorbed on surfaces, etc. However, when a system departs from the independent-particle approximation with strongly coherent photoemission and intrinsic secondary-electron emission, it requires the photoemission and the intrinsic secondary-electron (Auger- or Coster–Kronig (CK)-electron) emission to be examined with unprecedented discrimination. Under these conditions, measurement of the photoelectron and the intrinsic secondary electron in time coincidence can dramatically improve the understanding of the spectrum of either the photoemission or the intrinsic secondary-electron emission. By Auger–photoelectron coincidence spectroscopy (APECS), a photoelectron and its associated intrinsic secondary electron are detected in time coincidence. Since both of these electrons are generated by the respective creation and annihilation of the same initial core-hole state, only features characteristic of that event contribute to the coincidence spectrum. By explicitly exploiting this unique capability, APECS enables the photoemission and the intrinsic secondary-electron emission to be examined with unprecedented discrimination [1–21]. In this chapter we call the intrinsic secondary-electron spectrum simply the secondary-electron spectrum.

Recently APECS has been employed to analyze the singles (noncoincidence) secondary-electron (Auger-electron) spectra of metals [4–9]. When the photoelectron spectral line measured in coincidence with the selected singles secondary-electron spectral line, is compared with the singles photoelectron (X-ray photoelectron spectroscopy (XPS)) spectral line, the coincidence photoelectron spectral line shows an unexpected energy shift and asymmetrical narrowing compared to the singles photoelectron spectral line [8, 9]. The spectral behavior

Correlation Spectroscopy of Surfaces, Thin Films, and Nanostructures. Edited by Jamal Berakdar, Jürgen Kirschner
Copyright © 2004 Wiley-VCH Verlag GmbH & Co. KGaA, Weinheim
ISBN: 3-527-40477-5

remains puzzling so long as we maintain that we can interprete the coincidence spectrum in the same manner as we do the singles spectrum. In this chapter we show how to interpret the APECS spectra. At the same time we show that by manipulating the coincidence measurements we are somehow able to select a dynamical quantity of interest so that it predominantly governs the coincidence spectrum The puzzling feature is a result of such a selection. As an example, we discuss the many-body effect on the coincidence spectrum by the core-hole shakeup/down excitation, which leads to such puzzling spectral behavior. When one employs APECS for analysis of the singles spectra, it is absolutely necessary to have knowledge of the many-body effects inherent in the coincidence measurement, which may not manifest at all in the singles measurement.

12.2   APECS Spectrum

When the fixed incident photon energy, ω, is far above the ionization limit of a core-level electron of interest, the probability of observing the photoelectron (primary electron) and the secondary (Auger) electron, the kinetic energy of which are ϵ and ϵ_A, respectively, is [2, 10–12].

$$I(\epsilon, \epsilon_A)_\omega \sim \sum_{i,f} |Z(\epsilon)|^2 \, A(\epsilon - \omega) \left[\frac{\pi |V_{if}(\epsilon_A)|^2}{\operatorname{Im} \Sigma(\epsilon - \omega)} \right] D_f(\epsilon - \omega + \epsilon_A) \tag{12.1}$$

Z is the dipole-transition matrix element of the initial core-level electron photoionization. V_{if} is the effective secondary-electron emission matrix element which couples the initial core-hole state i with the final state f. D_f is the density of final states. A is the initial core-hole spectral function.

$$A(\epsilon - \omega) = \frac{1}{\pi} \frac{\operatorname{Im} \Sigma}{(\epsilon - \omega - \epsilon_c - \operatorname{Re} \Sigma)^2 + (\operatorname{Im} \Sigma)^2} \tag{12.2}$$

Here ϵ_c is the unperturbed initial core-hole energy. $\operatorname{Re} \Sigma$ and $\operatorname{Im} \Sigma$ are the real part and the imaginary part of the initial core-hole self-energy, respectively. They govern the energy shifts (spectral intensity) and widths of the initial core-hole states, respectively. When we integrate Eq. (12.1) over ϵ_A and sum over all decay channels, we obtain the singles photoelectron spectrum.

$$I(\epsilon)_\omega \sim |Z(\epsilon)|^2 \, A(\epsilon - \omega) \tag{12.3}$$

When we integrate Eq. (12.1) over ϵ, we obtain the singles secondary-electron spectrum

$$I(\epsilon_A)_\omega \sim \sum_{i,f} \int |Z(\epsilon)|^2 \, A(\epsilon - \omega) \left[\frac{\pi |V_{if}(\epsilon_A)|^2}{\operatorname{Im} \Sigma(\epsilon - \omega)} \right] D_f(\epsilon - \omega + \epsilon_A) \, d\epsilon \tag{12.4}$$

When the final state is localized and the configuration interaction including the lifetime effect is negligible. D_f can be approximated by a delta function. Then Eq. (12.4) becomes

$$I(\epsilon_A)_\omega \sim \sum_{i,f} |Z(\epsilon_f - \epsilon_A + \omega)|^2 \, A(\epsilon_f - \epsilon_A) \left[\frac{\pi |V_{if}(\epsilon_A)|^2}{\operatorname{Im} \Sigma(\epsilon_f - \epsilon_A)} \right] \tag{12.5}$$

Here ϵ_f is the final-state energy. The singles secondary-electron spectrum essentially consists of the convolution of D_f, and A multiplied by the branching ratio, i.e. the effective secondary-electron emission rate $(\pi|V_{if}|^2)$ divided by $\text{Im }\Sigma$.

When the secondary electron is measured in coincidence with the photoelectron, the kinetic energy of which is fixed to $\tilde{\epsilon}$, the coincidence secondary-electron spectrum is

$$I(\epsilon_A, \tilde{\epsilon})_\omega \sim \sum_{i,f} |Z(\tilde{\epsilon})|^2 A(\tilde{\epsilon} - \omega) \left[\frac{\pi |V_{if}(\epsilon_A)|^2}{\text{Im }\Sigma(\tilde{\epsilon} - \omega)} \right] D_f(\tilde{\epsilon} - \omega + \epsilon_A) \qquad (12.6)$$

When D_f can be approximated by a delta function, Eq. (12.6) becomes

$$I(\epsilon_A, \tilde{\epsilon})_\omega \sim \sum_{i,f} |Z(\tilde{\epsilon})|^2 A(\tilde{\epsilon} - \omega) \left[\frac{\pi |V_{if}(\epsilon_A)|^2}{\text{Im }\Sigma(\tilde{\epsilon} - \omega)} \right] \delta(\tilde{\epsilon} - \omega + \epsilon_A - \epsilon_f) \qquad (12.7)$$

The coincidence secondary-electron spectrum essentially consists of the product of the effective secondary-electron emission rate and D_f. Thus the lifetime broadening of the selected initial core-hole state does not contribute to the spectrum. When V_{if} is independent of the secondary-electron kinetic energy, the spectrum consists of D_f (Eq. (12.6)). Then when we shift the photoelectron analyzer by $\delta\epsilon$ from $\tilde{\epsilon}$ the coincidence secondary-electron line shifts by $-\delta\epsilon$ and the lineshape does not change. This is a consequence of the energy conservation. However, when V_{if} depends greatly on the secondary-electron kinetic energy, if we shift the photoelectron analyzer by $\delta\epsilon$ from $\tilde{\epsilon}$, the coincidence secondary-electron line shifts by $-\delta\epsilon$ but at the same time the lineshape changes. Thus by shifting the photoelectron analyzer, we can examine whether the secondary-electron emission rate depends considerably on the secondary-electron kinetic energy.

When the photoelectron is measured in coincidence with the secondary electron, the kinetic energy of which is fixed to $\tilde{\epsilon}_A$, the coincidence photoelectron spectrum is

$$I(\epsilon, \tilde{\epsilon}_A)_\omega \sim \sum_{i,f} |Z(\epsilon)|^2 A(\epsilon - \omega) \left[\frac{\pi |V_{if}(\tilde{\epsilon}_A)|^2}{\text{Im }\Sigma(\epsilon - \omega)} \right] D_f(\epsilon - \omega + \tilde{\epsilon}_A) \qquad (12.8)$$

When D_f can be approximated by a delta function, Eq. (12.8) becomes

$$I(\epsilon, \tilde{\epsilon}_A)_\omega \sim \sum_{i,f} |Z(\epsilon)|^2 A(\epsilon - \omega) \left[\frac{\pi |V_{if}(\tilde{\epsilon}_A)|^2}{\text{Im }\Sigma(\epsilon - \omega)} \right] \delta(\epsilon - \omega + \tilde{\epsilon}_A - \epsilon_f) \qquad (12.9)$$

The coincidence photoelectron spectrum is essentially the product of A and D_f, divided by $\text{Im }\Sigma$. If $\text{Im }\Sigma$ is independent of the hole energy $(= \epsilon - \omega)$, A becomes a Lorentzian. Then the spectrum is essentially the product of A and D_f. Thus when the width of D_f is smaller than that of A, the coincidence photoelectron spectrum becomes narrow compared to that of the singles photoelectron spectrum. The M_{23} photoelectron line of Cu(100) measured in coincidence with the singles M_{23}-VV (^1G) Auger-electron line is narrow compared to the singles photoelectron line [3]. We refer to Ref. [14] for a theoretical analysis of the APECS spectrum. When $\text{Im }\Sigma$ depends greatly on the hole energy, A becomes asymmetrical. In such a case we select a decay channel, V_{if} which is independent of the secondary-electron kinetic energy.

Then we can determine D_f from the coincidence secondary-electron spectrum (Eq. (12.6)). We measure the photoelectron in coincidence with the secondary electron emitted by the same decay channel (Eq. (12.8)). As A is known (Eq. (12.3)), we can determine Im Σ. Thus when the decay rate of interest is independent of the secondary-electron kinetic energy, a comparison of the singles photoelectron spectrum with the coincidence photoelectron spectrum can reveal the hole-energy dependence of the imaginary part of the initial core-hole self-energy.

Let us consider the case when an initial core-hole state decays to an intermediate two-hole (jk) state $|m\rangle$ by the CK transition. j is a core hole and k is a valence hole. The core-hole lifetime is much shorter than the valence-hole one so that the Auger transitions of the core-hole dominate for the decay of $|m\rangle$. When the valence-hole remains localized at the ionized atomic site during the core-hole decay and acts as a spectator hole, $|m\rangle$ is a two-hole state and the final state $|f\rangle$ is a three-hole state. However, when the valence-hole hops away rapidly from the ionized atomic site before the core-hole decays, $|m\rangle$ is a single core-hole state and $|f\rangle$ is a two-hole state. Then the cascade decay spectrum is governed by the hole dynamics of $|m\rangle$ (and $|f\rangle$). By measuring the Auger electron emitted by the core-hole decay of $|m\rangle$ in coincidence with the primary photoelectron, we can determine whether the valence hole localizes or delocalizes in time scale of core-hole decay.

When the Auger electron is measured in coincidence with the photoelectron, the kinetic energy of which is fixed to $\tilde{\epsilon}$, the coincidence Auger-electron spectrum is [10, 11, 18, 19]

$$
\begin{aligned}
I(\epsilon_A, \tilde{\epsilon})_\omega \sim \sum_{m,f} \int |Z(\tilde{\epsilon})|^2 \, A(\tilde{\epsilon} - \omega) \left[\frac{\pi |V_{im}(\epsilon_{CK})|^2}{\text{Im } \Sigma(\tilde{\epsilon} - \omega)} \right] D_m(\tilde{\epsilon} - \omega + \epsilon_{CK}) \\
\times \left[\frac{\pi |V_{mf}(\epsilon_A)|^2}{\Delta(\tilde{\epsilon} - \omega + \epsilon_{CK})} \right] D_f(\tilde{\epsilon} + \epsilon_{CK} + \epsilon_A - \omega) d\epsilon_{CK}
\end{aligned}
\tag{12.10}
$$

ϵ_A and ϵ_{CK} are the kinetic energies of the Auger electron and the CK-electron, respectively. D_m is the density of $|m\rangle$. Δ is the imaginary part of the complex energy-correction of $|m\rangle$. When the valence-hole acts as a spectator hole during the core-hole decay, it may affect the core-hole self-energy. Then Δ cannot be simply the sum of the imaginary part of the core-hole self-energy and that of valence-hole one. V_{im} is the CK-transition matrix element, while V_{mf} is the Auger-transition one. When V_{im} is energy independent in the kinetic energy region of interest and the valence-hole remains localized during the core-hole decay, the spectrum is the singles AES spectrum measured as if the core hole decayed as an "initial hole" in the presence of the valence-hole. The lifetime broadening of the selected initial state does not contribute to the coincidence spectrum. When V_{im} is energy independent but the valence-hole becomes delocalized before the core-hole decay, then the spectrum is the singles AES spectrum measured as if the core hole decayed as an "initial hole" without the presence of the valence hole. The APECS spectrum is governed by the valence-hole dynamics of $|m\rangle$ and $|f\rangle$. Thus compared to the singles measurement, by APECS we can know directly whether the valence hole remains localized or becomes delocalized in time scale of core-hole decay. We can determine the screening time of the valence hole in the presence of the core hole. Recently the author analyzed the APECS spectra of Pd, Sn and the late 3d-transition metals by a many-body theory [18, 19, 21].

When the photoelectron is measured in coincidence with the Auger electron, the kinetic energy of which is fixed to $\tilde{\epsilon}_A$, the coincidence photoelectron spectrum is [17]

$$I(\epsilon, \tilde{\epsilon}_A)_\omega \sim \sum_{m,f} \int |Z(\epsilon)|^2 \, A(\epsilon - \omega) \left[\frac{\pi |V_{im}(\epsilon_{CK})|^2}{\text{Im}\, \Sigma(\epsilon - \omega)} \right] D_m(\epsilon - \omega + \epsilon_{CK})$$

$$\times \left[\frac{\pi |V_{mf}(\tilde{\epsilon}_A)|^2}{\Delta(\epsilon - \omega + \epsilon_{CK})} \right] D_f(\epsilon + \epsilon_{CK} + \tilde{\epsilon}_A - \omega) \, d\epsilon_{CK}$$

$$(12.11)$$

When the final state is not interacting and the lifetime broadening is negligible, the spectrum is

$$I(\epsilon, \tilde{\epsilon}_A)_\omega \sim \sum_{m,f} |Z(\epsilon)|^2 \, A(\epsilon - \omega) \left[\frac{\pi |V_{im}(\epsilon_f + \omega - \tilde{\epsilon}_A - \epsilon)|^2}{\text{Im}\, \Sigma(\epsilon - \omega)} \right] D_m(\epsilon_f - \tilde{\epsilon}_A)$$

$$\times \left[\frac{\pi |V_{mf}(\tilde{\epsilon}_A)|^2}{\Delta(\epsilon_f - \tilde{\epsilon}_A)} \right]$$

$$(12.12)$$

As both D_m and the factor inside the second square brackets (i.e., the branching ratio of the Auger-transition rate of core-hole j) are energy independent, the spectrum is governed by the product of A and the factor inside the first square brackets, i.e., the branching ratio of the CK-transition rate of the initial core-hole. Thus the branching ratio can be determined by the singles photoelectron spectrum (A) and the coincidence one. When V_{im} is considerably energy-dependent, the coincidence photoelectron line can be shifted and distorted compared to the singles one, unless the CK transition dominates in the initial core-hole decay. Recently the author analyzed the puzzling energy shift of the L_2 photoelectron line of Cu metal measured in coincidence with the singles L_2-L_3V - VVV AES spectral line by a many-body theory [17].

12.3   Shakeup/down and Coincidence Photoelectron Spectrum.

From the discussion in Section 12.2, it is clear that either when the initial core-hole self-energy, particularly the imaginary part of which, depends critically on the hole energy, or when the partial branching ratio is energy dependent, the coincidence photoelectron spectrum may behave considerably differently from the singles one. As such an example, recently a pronounced interference effect in the APECS spectrum is predicted, when the core-hole decay develops along two different paths, i.e. by a direct Auger-transition path and by an indirect path passing via the two-hole one-particle excitation and de-excitation. The screening of the direct Auger-transition matrix element by the particle-hole (de)excitations leads to a pronounced interference effect in the decay spectrum [20]. In this section, as another example we show how the initial core-hole shakeup/down affects the coincidence photoelectron spectrum compared to the singles one We consider the self-energy of the initial core-hole state interacting with the two-hole one-particle (2h,1p) state by the shakeup/down. We study how the imaginary part of the self-energy, i.e. the indirect decay of the initial core-hole state by the configuration interaction with the 2h,1p state which subsequently decays, affects the coincidence photoelectron spectrum compared to the singles one. We consider two unperturbed discrete core-hole states $|1\rangle$ and $|2\rangle$ coupled by the shakeup/down. $|1\rangle$ is the single core-hole state, while $|2\rangle$ is the 2h,1p state. In the latter an empty discrete-level is occupied by an electron from an occupied discrete-level. For the sake of simplicity we neglect the band structures and the multiplet splittings of the 2h,1p state. Such an assumption does not affect the physics discussed here.

The initial core-hole propagator of $|1\rangle$ is

$$G_1(\epsilon - \omega) = (\epsilon - \omega - \epsilon_1 - \Sigma(\epsilon - \omega) - i\Gamma_1)^{-1} \tag{12.13}$$

ϵ_1, is the unperturbed core-hole energy of $|1\rangle$. Σ is the core-hole self-energy of $|1\rangle$ by the aforementioned interaction with $|2\rangle$. Γ_1 is the core-hole decay width of $|1\rangle$ by the direct decay of $|1\rangle$. Γ_1 is energy independent. We neglect the core-hole energy shift by the corresponding virtual Auger transitions because it is very small. Then the core-hole spectral function is predominantly governed by the interaction between $|1\rangle$ and $|2\rangle$. The self-energy is

$$\Sigma(\epsilon - \omega) = V^2(\epsilon - \omega - \epsilon_2 - i\Gamma_2)^{-1} \tag{12.14}$$

Here ϵ_2 is the unperturbed 2h,1p state energy of $|2\rangle$. V is the energy-independent shakeup/down matrix element by which $|1\rangle$ is coupled with $|2\rangle$. Γ_2 is the core-hole decay width of $|2\rangle$. The decay widths of $|1\rangle$ and $|2\rangle$ are the sum over the decay channel of $|1\rangle$ and that of $|2\rangle$, respectively.

$$\Gamma_1 = \sum_i \Gamma_1^i \tag{12.15}$$

$$\Gamma_2 = \sum_j \Gamma_2^j \tag{12.16}$$

The final two-hole state $|f_1^i\rangle$ by the core-hole decay channel i of $|1\rangle$ is assumed to be well separated from the final three-hole one-particle state $|f_2^j\rangle$ by the core-hole decay channel j of $|2\rangle$ so that the interaction between $|f_1^i\rangle$ and $|f_2^j\rangle$ is negligible. Thus the interference between $|1\rangle$ and $|2\rangle$ through the final states is negligible. Moreover, $|f_1^i\rangle$ and $|f_2^j\rangle$ are assumed to be localized and their lifetime broadening is negligible. Re Σ governs the energy shift (spectral intensity) of the initial core-hole state $|1\rangle$, while Im Σ gives the decay width of $|1\rangle$ by the indirect decay of $|1\rangle$ via the interaction with $|2\rangle$, which decays to $|f_2^j\rangle$ (see Figure 12.1).

The core-hole spectral function $A(\epsilon - \omega)$ or the singles photoelectron spectrum, is given by the imaginary part of Eq. (12.13) divided by π

$$A(\epsilon - \omega) = \frac{1}{\pi} \frac{\Gamma_1 + \mathrm{Im}\, \Sigma(\epsilon - \omega)}{(\epsilon - \omega - \epsilon_1 - \mathrm{Re}\, \Sigma(\epsilon - \omega))^2 + (\Gamma_1 + \mathrm{Im}\, \Sigma(\epsilon - \omega))^2} \tag{12.17}$$

We decompose A into two parts A_1 and A_2.

$$A_1(\epsilon - \omega) = A(\epsilon - \omega) \left[\frac{\Gamma_1}{\Gamma_1 + \mathrm{Im}\, \Sigma(\epsilon - \omega)} \right] \tag{12.18}$$

$$A_2(\epsilon - \omega) = A(\epsilon - \omega) \left[\frac{\mathrm{Im}\Sigma(\epsilon - \omega)}{\Gamma_1 + \mathrm{Im}\, \Sigma(\epsilon - \omega)} \right] \tag{12.19}$$

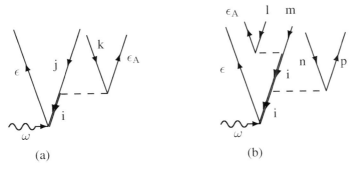

Figure 12.1: The open diagrams for the Auger transitions of the initial core-hole state discussed in the text. (a) Direct Auger transition of $|1\rangle$ to $|f_1^i\rangle$. Here $|1\rangle$ is the single core-hole state $\underline{i}$, while $|f_1^i\rangle$ is the two-hole state $\underline{jk}$. The underbar denotes the hole state. ϵ is the primary photoelectron kinetic energy and ω is the incident photon energy. ϵ_A is the Auger-electron kinetic energy. The initial core-hole propagator in the diagram is dressed by the self-energy (Σ) of $|1\rangle$ interacting with the two-hole one-particle (2h,1p) state $|2\rangle$ ($\underline{in}p$) by the shakeup/down excitation, which subsequently decays to the three-hole one-particle (3h,1p) state $|f_2^j\rangle$ ($\underline{lmn}p$). The propagator is dressed also by the (imaginary part of) self-energy by the direct Auger transitions of $|1\rangle$ to $\sum_i |f_1^i\rangle$, i.e. Γ_1. (b) Indirect Auger transition of $|1\rangle$ via the interaction with the 2h,1p state $|2\rangle$ by the shakeup/down excitation, which subsequently decays to the 3h,1p state $|f_2^j\rangle$. The lower part of the core-hole propagator in the diagram is dressed in the same manner as that in (a), while the upper one is dressed by the (imaginary part of) self-energy by the Auger transitions $|2\rangle$ to $\sum_j |f_2^j\rangle$, i.e. Γ_2. Note that indices i and j in $|f_1^i\rangle$ and $|f_2^j\rangle$ are *not* the hole indices i and j in the diagrams.

A_1 is the part of A due to the direct decay of $|1\rangle$, while A_2 is that due to the indirect decay of $|1\rangle$ by the interaction with $|2\rangle$ which subsequently decays.

The singles secondary-electron spectrum by direct decay channel i of $|1\rangle$ and the one by indirect decay of $|1\rangle$ by the interaction with $|2\rangle$ which subsequently decays by decay channel j, are [10, 11]

$$I(\epsilon_A)_\omega \sim A_1(\epsilon_{f_1}^i - \epsilon_A)\frac{\Gamma_1^i}{\Gamma_1} + A_2(\epsilon_{f_2}^j - \epsilon_A)\frac{\Gamma_2^j}{\Gamma_2} \tag{12.20}$$

Here $\epsilon_{f_1}^i$ and $\epsilon_{f_2}^j$ are the energy of the final state by decay channel i of $|1\rangle$ and that by decay channel j of $|2\rangle$, respectively.

We measure the photoelectron in coincidence with the secondary-electron line, the kinetic energy of which is ϵ_A^0. Then the coincidence photoelectron spectrum is [10, 11]

$$I(\epsilon, \epsilon_A^0)_\omega \sim A_1(\epsilon - \omega)\frac{\Gamma_1^i}{\Gamma_1}\delta(\epsilon - \omega - \epsilon_{f_1}^i + \epsilon_A^0)$$
$$+ A_2(\epsilon - \omega)\frac{\Gamma_2^j}{\Gamma_2}\delta(\epsilon - \omega - \epsilon_{f_2}^j + \epsilon_A^0) \tag{12.21}$$

The coincidence-photoelectron consists of $A_1^i = A_1(\Gamma_1^i/\Gamma_1)$ at $\epsilon - \omega = \epsilon_{f_1}^i - \epsilon_A^0$ by the direct decay of $|1\rangle$ and $A_2^j = A_2(\Gamma_2^j/\Gamma_2)$ at $\epsilon - \omega = \epsilon_{f_2}^j - \epsilon_A^0$ by the indirect decay of $|1\rangle$. When in the singles secondary-electron spectrum $A_1^i(\epsilon_{f_1}^i - \epsilon_A)$ and $A_2^j(\epsilon_{f_2}^j - \epsilon_A)$ are well separated so that ϵ_A^0 corresponds only to the singles secondary-electron line A_1^i (or A_2^j), the coincidence photoelectron spectrum consists only of the first term (or the second term).

When the final-state lifetime broadening is not negligible, the coincidence photoelectron spectrum becomes

$$
\begin{aligned}
I(\epsilon, \epsilon_A^0)_\omega \sim\ & A_1(\epsilon - \omega) \frac{\Gamma_1^i}{\Gamma_1} D_{f_1}^i (\epsilon - \omega + \epsilon_A^0) \\
& + A_2(\epsilon - \omega) \frac{\Gamma_2^j}{\Gamma_2} D_{f_2}^j (\epsilon - \omega + \epsilon_A^0)
\end{aligned}
\tag{12.22}
$$

The coincidence-photoelectron spectrum consists of the product of A_1^i and $D_{f_1}^i$ and that of A_2^j and $D_{f_2}^j$. Here $D_{f_1}^i$ (or $D_{f_2}^j$) is the density of final states $|f_1^i\rangle$ (or $|f_2^j\rangle$), and $\Gamma_1(\Gamma_2)$ and $\Gamma_1^i(\Gamma_2^j)$ are defined by

$$
\Gamma_1(\epsilon - \omega) = \sum_i \pi \int \left| V_{1f_1}^i(\epsilon_A) \right|^2 D_{f_1}^i(\epsilon - \omega + \epsilon_A) \mathrm{d}\epsilon_A
\tag{12.23}
$$

$$
\Gamma_2(\epsilon - \omega) = \sum_j \pi \int \left| V_{2f_2}^j(\epsilon_A) \right|^2 D_{f_2}^j(\epsilon - \omega + \epsilon_A) \mathrm{d}\epsilon_A
\tag{12.24}
$$

$$
\Gamma_1^i = \pi \left| V_{1f_1}^i(\epsilon_A^0) \right|^2
\tag{12.25}
$$

$$
\Gamma_2^j = \pi \left| V_{2f_2}^j(\epsilon_A^0) \right|^2
\tag{12.26}
$$

Here $V_{1f_1}^i$ and $V_{2f_2}^j$ are the Auger transition matrix elements by which $|1\rangle$ and $|f_1^i\rangle$, $|2\rangle$ and $|f_2^j\rangle$ are coupled. In the following discussion we assume they are energy independent. From Eqs. (12.19)–(12.21)(12.24)(12.26) it is clear that the difference between the singles photo-electron spectrum and the coincidence one is due to $\mathrm{Im}\Sigma$.

In Figures 12.2 and 12.3 we show the initial core-hole spectral function $A(E)$ calculated for different $E_0, \Gamma_1 = \Gamma_2$ and $V = 1.0$ or $V = 2.0$. In Table 12.1 we list $E = E_{\max}$ at which the peaks of $A(E)$ appear. The singles photoelectron spectrum is essentially $A(E)$ where $E = \epsilon - \omega - \epsilon_2$ is the hole energy of $|1\rangle$ defined relative to the unperturbed energy of $|2\rangle$. $E_0 = \epsilon_1 - \epsilon_2$, is the unperturbed hole energy of $|1\rangle$ relative to the unperturbed energy of $|2\rangle$. $A(E)$ for $E_0 = -\xi$ is the mirror image of $A(E)$ for $E_0 = \xi$. The main-line state of larger spectral intensity is predominantly a single-hole state, while the satellite is predominantly a shakeup (or shakedown) 2h,1p state. With an increase in V, the mixing between $|1\rangle$ and $|2\rangle$ becomes significant. The spectral behavior essentially follows what is expected by a simple two-level model.

In Figures 12.2 and 12.3 we also show $A_1^i(E)$ and $A_2^j(E)$ calculated for the same set of parameter values used for $A(E)$. For illustration in Figure 12.2 and 12.3 we plot $A_1(E)$ and $A_2(E)$ instead of $A_1^i(E)$ and $A_2^j(E)$. Γ_1^i and Γ_2^j are energy independent. Thus $A_1^i(E)$

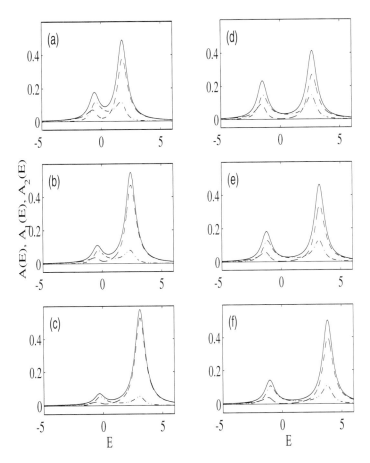

Figure 12.2: (a)–(c) The core-hole spectral function $A(E)$ (full curves) calculated for $V = 1.0$, $\Gamma_1 = \Gamma_2 = 0.5$ and from top to bottom $E_0 = 1.2$, 2.0 and 2.8 respectively. Also shown are $A_1(E)$ (dashed curves) and $A_2(E)$ (chain curves) for the same parameter values. (d)–(f) same as above except for $V = 2.0$. E is the initial hole energy (see the text for details).

(or $A_2^j(E)$) is proportional to A_1 (or A_2). The former lineshape is the same as that of the latter one. In Table 12.1 we list $E = E_{\max}$ at which the peaks of $A_1^i(E)$ (or $A_1(E)$) or $A_2^j(E)$ (or $A_2(E)$) appear. The single secondary-electron spectrum consists of $A_1^i(E)$ and $A_2^j(E)$ (Eq. (12.22)). For the singles secondary-electron spectrum $E = \epsilon_{f_1}^i - \epsilon_A - \epsilon_2$ for $A_1^i(E)$ and $E = \epsilon_{f_2}^j - \epsilon_A - \epsilon_2$ for $A_2^j(E)$. $A_1^i(E)$ is due to the direct decay of $|1\rangle$ by the Auger transition from $|1\rangle$ to $|f_1^i\rangle$, while $A_2^j(E)$ is due to the indirect decay of $|1\rangle$ by the interaction with $|2\rangle$ which subsequently decays to $|f_2^j\rangle$ by the Auger transition. A_1^i or A_2^j differs considerably from the singles photoelectron spectrum A. The Auger-electron kinetic energy of the direct decay of $|1\rangle$ to $|f_1^i\rangle$, i.e. $\epsilon_A^i = \epsilon_{f_1}^i - \epsilon + \omega$, often differs substantially from

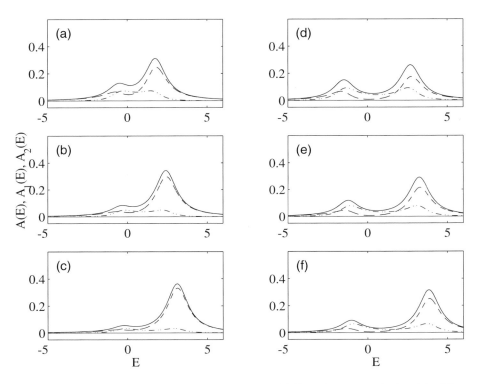

Figure 12.3: (a)–(c) The core-hole spectral function $A(E)$ (full curves) calculated for $V = 1.0$, $\Gamma_1 = \Gamma_2 = 0.8$ and from top to bottom $E_0 = 1.2, 2.0$ and 2.8 respectively. Also shown are $A_1(E)$ (dashed curves) and $A_2(E)$ (chain curves) for the same parameter values. (d)–(f) same as above except for $V = 2.0$. E is the initial hole energy (see the text for details)

that of the indirect decay of $|1\rangle$ to $|f_2^j\rangle$ via the interaction with $|2\rangle$, i.e. $\epsilon_A^j = \epsilon_{f_2}^j - \epsilon + \omega$. In this case, in the singles secondary-electron spectrum A_1^i is well separated from A_2^j Then the singles secondary-electron spectral peak of $A_1^i(E)$ (or $A_2^j(E)$) does not coincide with the singles photoelectron spectral peak of $A(E)$, when the latter spectrum is shifted by using the energy of $|f_1^i\rangle$ (or $|f_2^j\rangle$) (see Table 12.1). Even when the Auger electron kinetic energy of the direct decay of $|1\rangle$ is rather close to that of the indirect one so that in the singles secondary-electron spectrum A_1^i and A_2^j overlap, the aforementioned energy shift between A and the singles AES spectrum which consists of A_1^i and A_2^j, is expected. Because only when $\epsilon_{f_1}^i = \epsilon_{f_2}^j$, is A proportional to the sum of A_1^i and A_2^j as shown in Figures 12.2 and 12.3. When the satellite is a shakeup (or shakedown) state, the singles secondary-electron spectral line by the direct decay of $|1\rangle$, i.e. $A_1^i(E)$, shifts toward the smaller (or larger) secondary-electron kinetic energy from $\epsilon_A = \epsilon_{f_1}^i - \epsilon_c$. Here ϵ_c is the main-line state core-hole energy, i.e. the core-hole energy corresponding to the singles photoelectron main-line peak. When A_1^i and A_2^j in the singles secondary-electron spectrum are well separated the linewidth of the singles

Table 12.1: Hole energy $E_{\max}$ at which the peak of $A(E)$, $A_1(E)$ or $A_2(E)$ appears.

	$\Gamma_1 = \Gamma_2 = 0.5, V = 1.0$						$\Gamma_1 = \Gamma_2 = 0.5, V = 2.0$					
E_0	$A(E)$		$A_1(E)$		$A_2(E)$		$A(E)$		$A_1(E)$		$A_2(E)$	
1.2	−0.55	1.76	−0.72	1.79	−0.45	1.65	−1.49	2.69	−1.58	2.72	−1.43	2.63
2.0	−0.40	2.41	−0.60	2.43	−0.32	2.32	−1.23	3.24	−1.35	3.26	−1.18	3.18
2.8	−0.31	3.12	−0.54	3.13	−0.25	3.05	−1.04	3.84	−1.18	3.85	−0.99	3.79
	$\Gamma_1 = \Gamma_2 = 0.8, V = 1.0$						$\Gamma_1 = \Gamma_2 = 0.8, V = 2.0$					
E_0	$A(E)$		$A_1(E)$		$A_2(E)$		$A(E)$		$A_1(E)$		$A_2(E)$	
1.2	−0.47	1.76	−0.74	1.82	−0.25	1.45	−1.48	2.69	−1.67	2.76	−1.33	2.53
2.0	−0.31	2.41	−0.62	2.44	−0.17	2.17	−1.22	3.23	−1.46	3.28	−1.09	3.09
2.8	−0.22	3.12	−0.54	3.14	−0.12	2.92	−1.03	3.84	−1.3	3.87	−0.91	3.71

secondary-electron spectrum (A_1^i or A_2^j) becomes narrow compared with that of the singles photoelectron spectrum.

The reason why the singles photoelectron line peak does not coincide with the singles secondary-electron spectrum, when the latter is shifted by the exact final-state energy, is the following. The degree of the change in the lineshape such as the line shift and asymmetrical narrowing from A to A_1 depends on the branching ratio of A_1 in Eq. (12.20) , i.e., the ratio of the energy-independent part of the imaginary part of the self-energy, i.e., Γ_1 relative to the total imaginary part of the self-energy. The numerator of the branching ratio of A_1 is energy independent so that the change in the lineshape from A to A_1 depends on the ratio of Im Σ relative to Γ_1. Im Σ is a Lorentzian which maximizes at the unperturbed 2h,1p state energy, i.e., $E = 0$. Thus the branching ratio of A_1 minimizes at $E = 0$ so that A_1 minimalizes at $E = 0$. ImΣ decreases as the hole energy decreases or increases from $E = 0$ so that the branching ratio of A_1 increases and approaches to 1.0. Thus the spectral intensity of A_1 is much reduced around $E = 0$ compared to that of A but the former becomes less reduced compared to the latter as the hole energy decreases or increases from $E = 0$. This explains the line shift and narrowing of A_1 compared to A shown in Figure 12.2 and 12.3. Note that the change of ImΣ depends on the decay width of $|2\rangle$, Γ_2. Then with an increase in Γ_2, the line shift increases (see Table 12.1).

The degree of the change in the lineshape from A to A_2 depends on the branching ratio of A_2 in Eq. (12.21), i.e., the ratio of Im Σ relative to the total imaginary part of the self-energy. As Im Σ in the branching ratio of A_2 maximizes at $E = 0$, the branching ratio of A_2 maximizes at $E = 0$ so that the difference between A_2 and A minimizes at $E = 0$. Im Σ decreases as the hole energy decreases or increases from $E = 0$ so that the branching ratio of A_2 decreases at the same time. Thus as the hole energy decreases or increases from $E = 0$, the spectral intensity of A_2 becomes more reduced compared to A. This explains the line shift and narrowing of A_2 compared to A shown in Figures 12.2 and 12.3. It is the initial core-hole self-energy by the shakeup/down excitation between the initial core-hole state and the 2h,1p state, which plays a key role.

It should be noted that we made an assumption that $\Gamma_1 = \Gamma_2$. The decay rate of the core-hole in the presence of the valence hole created by the core-hole shakeup/off excitation should be similar to that of the core-hole without the presence of the valence hole, if the core-hole decay is predominantly governed by the Auger transitions. Because the Auger transition rate is fairly independent of the final-state potential (two holes or three holes) and the change in the Auger-electron kinetic energy by the presence of the extra "spectator" hole. Thus $\Gamma_1 = \Gamma_2$ is often valid.

First we consider the coincidence photoelectron spectrum when in the singles secondary-electron spectrum A_1^i and A_2^j of Eq. (12.22) are well separated. We define the secondary-electron kinetic energy of the peak of the singles secondary-electron spectrum A_1^i (or A_2^j) as $\epsilon_A(1)$ (or $\epsilon_A(2)$). Then when the final-state interaction and the lifetime broadening are negligible, if we measure the photoelectron in coincidence with the peak of the singles secondary-electron spectrum A_1^i , by Eq. (12.24) we observe the coincidence photoelectron line at $\epsilon(1) = \omega + \epsilon_{f_1}^i - \epsilon_A(1)$ with the intensity of A_1^i at $\epsilon = \epsilon(1) = E_{max} + \omega$. E_{max} is that of $A_1^i(A_1)$. As shown in Table 12.1 E_{max} of $A_1^i(A_1)$ does not coincide with that of A (singles photoelectron spectrum). If we measure the photoelectron in coincidence with the peak of the singles secondary-electron spectrum A_2^j , we observe the coincidence photoelectron line at $\epsilon(2) = \omega + \epsilon_{f_2}^j - \epsilon_A(2)$ with the intensity A_2^j at $\epsilon = \epsilon(2) = E_{max} + \omega$. E_{max} is that of A_2^j. E_{max} of $A_2^j(A_2)$ does not coincide with that of A. Thus the coincidence photoelectron line peak does not coincide with the singles photoelectron one. When A_1^i and A_2^j overlap in the singles secondary-electron spectrum, in the former case we observe an extra coincidence photoelectron line at $\epsilon'(1) = \omega + \epsilon_{f_2}^j - \epsilon_A(1)$ with the intensity of A_2^j at $\epsilon = \epsilon'(1)$, while in the latter case we observe an extra one at $\epsilon'(2) = \omega + \epsilon_{f_1} - \epsilon_A(2)$ with the intensity of A_1^i at $\epsilon = \epsilon'(2)$ (see Eq. (12.24)). When the final-state lifetime broadening is not negligible, the coincidence photoelectron spectrum consists of the product of A_1^i (or A_2^j) with $D_{f_1}^i$ (or $D_{f_2}^j$) evaluated for the kinetic energy of the secondary-electron of interest.

The coincidence photoelectron peak does not necessarily coincide with the singles photoelectron peak. Whether the former coincides with the latter can be examined only by the coincidence measurement, unless the exact two-hole energy is known. The "exact" two-hole energy is often determined by using the singles photoelectron line peak-energy and the singles secondary-electron line peak-energy. However, when the initial core-hole state is interacting, the coincidence photoelectron line measured in coincidence with the singles secondary-electron line does not necessarily coincide with the singles photoelectron one. Thus in such a case the "exact" two-hole energy determined by the singles secondary-electron line energy and the singles photoelectron one is not the exact two-hole energy.

12.4 Coincidence L_3 Photoelectron Line of Cu Metal

The L_3 photoelectron line of Cu metal measured in coincidence with the singles L_3 - VV (^{1}G) Auger-electron line is narrowed on the low kinetic energy side and shifts by 0.25 eV to the high kinetic energy side of the singles photoelectron line. However, the L_3 photoelectron line measured in coincidence with the singles L_3 - VV (^{3}F) Auger-electron line, shifts by 0.15 eV to the low kinetic energy side of the singles photoelectron line peak [8].

The L_3 - VV Auger-electron spectrum of Cu metal measured in coincidence with the singles L_2 photoelectron main line, shows the presence of the satellite due to the $L_2 - L_3$ V

- VVV transition [4–6]. The major part of the satellite is about 2.5 eV below the L_3 - VV (^{1}G) Auger-electron line. However, there is also a satellite of considerable intensity around the ^{1}G line. The L_3 photoelectron spectrum of Cu metal measured in coincidence with the satellite about 2.5 eV below the L_3 - VV (^{1}G) Auger-electron line shows that the L_3 V - VVV transition is as substantial as the $L_2 - L_3$ V - VVV transition [6].

When the secondary-electron analyzer is set on the ^{3}F line, the probability that the analyzer collects the secondary-electron emitted by the L_3 V - VVV decay of the L_3V shakeup/off state is not small because there is a satellite of substantial intensity around the ^{3}F line. Then the probability that the photoelectron analyzer collects the photoelectron by the L_3 V shakeup/off in coincidence with the ^{3}F line, is not small. In other words the coincidence L_3 photoelectron main line consists of the product of the density of the final states and $A_1(E)$ weighted by the branching ratio of the L_3 - VV (^{3}F) Auger-transition rate, and the product of the density of the final states and $A_2(E)$ weighted by the branching ratio of the L_3V - VVV transition rate (Eq. (12.26)). As $A_2(E)$ is centered on the low kinetic energy side of the singles photoelectron main line (see $A_2(E)$ in Figures 12.2 and 12.3), the coincidence L_3 photoelectron main line shifts toward the low kinetic energy side of the singles one.

There is also a satellite of substantial intensity at the ^{1}G line [6]. Thus when the secondary-electron analyzer is set on the ^{1}G line, the probability that the photoelectron analyzer collects the photoelectron by the L_3 V shakeup/off is not small. Then the coincidence L_3 photoelectron main line consists of the product of the density of the final states and $A_1(E)$ weighted by the branching ratio of the L_3-VV (^{1}G) Auger-transition rate, and the product of the density of the final states and $A_2(E)$ weighted by the branching ratio of the L_3V - VVV transition rate (Eq. (12.26)). However, the branching ratio of the L_3-VV (^{1}G) Auger-transition rate is much larger than that of the L_3 - VV (^{3}F) Auger-transition rate so that $A_1(E)$ weighted by the branching ratio of the L_3 - VV (^{1}G) Auger-transition rate, dominates the coincidence L_3 photoelectron main line. Then the coincidence photoelectron main line is narrowed on the low kinetic energy side and shifts to the high kinetic energy side of the singles one (see $A_1(E)$ in Figures 12.2 and 12.3).

The coincidence L_3 photoelectron line of Cu metal was found to be unshifted, when the secondary-electron analyzer was shifted by 0.7 or 1.0 eV from the singles secondary-electron main-line [8]. For the sake of simplicity we assume that both initial core-hole spectral function, A, and the density of final two-hole states, D_f, in Eq. (12.8) are Lorentzians. Then using the experimental or semiempirical lifetime widths of the initial L_3 core-hole state and final two 3d-hole states of Cu metal. we predict that when the secondary-electron analyzer is set 0.7 eV below (or above) the singles L_3-VV Auger-electron line, the coincidence L_3 photoelectron line shifts by as little as 0.03 eV towards higher (or lower) kinetic energy. When the secondary-electron analyzer is set 1.0 eV below (or above) the singles L_3- VV Auger-electron line, the coincidence L_3 photoelectron line shifts also by 0.03 eV towards higher (or lower) kinetic energy. The energy shift is much smaller than the experimental resolution. Thus the coincidence L_3 photoelectron line of Cu metal remains unshifted, when the secondary-electron analyzer is shifted from the singles secondary-electron line

To see the effect of the density of final two-hole states on the coincidence L_3 photoelectron spectrum (Eq. (12.26)), we assume the aforementioned conditions for Eq. (12.8). Then we predict that the coincidence photoelectron line narrows by 0.03 eV compared to the singles photoelectron line. The narrowing is much smaller than the experimental resolution. Thus so long as the density of final states is a Lorentzian, i.e. the final state is a two-valence hole localized state broadened by the lifetime, the L_3 photoelectron line of Cu metal measured in coincidence with the singles L_3 - VV ^{1}G (or ^{3}F) Auger-electron line, should be essentially described by the aforementioned superposition of $A_1^i(E)$ and $A_2^j(E)$.

The direction of the energy shift and asymmetrical narrowing of the coincidence photoelectron main line compared to the singles one, depends not only on the initial core-hole shakeup/down satellite intensity but also on how large the multiple-hole satellite by the decay of the initial core-hole shakeup/off satellite is in the secondary-electron line, on which the secondary-electron analyzer is set. The theory solves the puzzling energy shift and asymmetrical narrowing of the coincidence L_3 photoelectron main line of Cu metal, which depends on which secondary-electron line the secondary-electron analyzer is set on. The theory predicts that when the shakeup/off satellite is negligible, the coincidence photoelectron main-line coincides with the singles one. The prediction is confirmed by the M_{45} photoelectron line of Ag metal measured in coincidence with the singles $M_{45} - N_{45}N_{45}$ secondary-electron line [9]. This study shows one of the unique capabilities of the APECS, which has not been perceived before.

12.5   Concluding Remarks

The coincidence photoelectron spectrum is the singles photoelectron spectrum divided by the imaginary part of the core-hole self-energy and multiplied by the density of final states. Thus when the imaginary part of the core-hole self-energy depends critically on the core-hole energy in the energy region of interest, the coincidence photoelectron spectrum may differ substantially from the singles one. Then the difference between the two spectra is nothing but the manifestation of the core-hole dynamics. As an example, the core-hole shakeup/down effect on the coincidence photoelectron spectrum is discussed. So far we have not considered the possibility of the relaxation of an incompletely relaxed core-hole state to a fully relaxed one in time scale of core-hole decay. When relaxation is possible, by APECS we can determine the relaxation time. Such a capability of APECS which has not been perceived yet is discussed in Ref. [12].

References

[1] H.W. Haak, G.A. Sawatzky. and T.D. Thomas, Phys. Rev. Lett. **41**, (1978) 1825.

[2] M. Ohno and G. Wendin, J. Phys. B**12**, (1979) 1305.

[3] E. Jensen, R.A. Bartynski, S.L. Hulbert, E.D. Johnson, and R. Garrett, Phys. Rev, Lett. **62**, (1989) 71.

[4] S.M. Thurgate, J. Electron. Spectrosc. Relat. Phenom. **77**, (1996) 281.

[5] S.M. Thurgate, J. Electron. Spectrosc. Relat. Phenom. **81**, (1996) 1 and references therein.

[6] S.M. Thurgate, C.P. Lund, C. Creagh, and R. Craig, J. Electron Spectroscopy and Relat. Phenom. **93**, (1998) 209.

[7] C. P. Lund and S.M. Thurgate, Surface Sci. **376** (1997) L403.

[8] Z.-T. Jiang, S.M. Thurgate, and P. Wilkie, Surf. Interface Anal. **31**, (2001) 287.

[9] S.M. Thurgate, in *Surface Analysis by Auger and X-ray Photoelectron Spectroscopy* Briggs and Grant eds., IM Publications, in press (2003).

[10] M. Ohno, Phys. Rev. B **58**, (1998-I) 12795.

[11] M. Ohno, Phys. Rev. B **59**, (1999-I) 10371.

[12] M. Ohno, J. Electron. Spectrosc. Relat. Phenom. **104**, (1999) 109.

[13] M. Ohno, J. Electron. Spectrosc. Relat. Phenom. **109**, (2000) 233.

[14] M. Ohno, J. Electron. Spectrosc. Relat. Phenom. **124**, (2002) 39.

[15] M. Ohno, J. Electron. Spectrosc. Relat. Phenom. **124**, (2002) 53.

[16] M. Ohno, J. Electron. Spectrosc. Relat. Phenom. **124**, (2002) 61.

[17] M. Ohno, J. Electron. Spectrosc. Relat. Phenom. **125**, (2002) 161.

[18] M. Ohno, J. Electron. Spectrosc. Relat. Phenom., (2003) in press.

[19] M. Ohno, J. Electron. Spectrosc. Relat. Phenom., (2003) in press.

[20] M. Ohno and L. Sjögren, J. Phys. B **36** (2003) 4519.

[21] M. Ohno, to be published.

13 Auger–Photoelectron Coincidence Spectroscopy (APECS) of Transition Metal Compounds

Robert A. Bartynski, Alex K. See, Wing-Kit Siu, and Steven L. Hulbert

Transition metal (TM) compounds exhibit a wide variety of interesting physical properties that are intimately tied to the d-level occupancy of the transition metal component. Owing to its site specificity and sensitivity to charge transfer, Auger–photoelectron coincidence spectroscopy (APECS) can be used to explore the electronic structure of transition metal compounds, and their interaction with other systems, in a unique way. To demonstrate these capabilities, we have used APECS to study the clean and metal-covered stoichiometric $TiO_2(110)$ surface. We have determined the intrinsic line shape of the Ti $M_{23}VV$ Auger transition and find that it consists of two components, which we associate with two decay channels: one where the two final-state holes propagate together to the same O-site, the other where the holes propagate to different O-sites. In addition, we use the extreme sensitivity of APECS to the presence of reduced Ti sites to determine the role of surface defects in the nucleation of Cu and Ag overlayers on the $TiO_2(110)$ surface. We find that Cu nucleates at sites in a Ti^{2+} oxidation state while Ag nucleates at Ti^{3+} sites. These results provide insight into recent STM studies of these systems.

13.1 Introduction

Transition metal compounds have garnered considerable interest in recent years, in part because, with relatively subtle changes in composition or stoichiometry, a single compound can exhibit a wide variety of interesting electronic, magnetic, or chemical properties. The versatility of these systems is attributed to a delicate balance between orbital, lattice, and spin degrees of freedom, all having very similar energy scales. In many cases, substitutional TM compounds (e.g., compounds of the form $A_xB_{1-x}MnO_3$, where A and B are column I and column II elements, respectively) exhibit their most interesting properties at relatively low impurity concentrations, rather than at the end point of the phase diagram where the TM ion is in its maxim valence state. This can be because the dopant will chemically reduce some of the TM ions, leading to interesting interactions among the ions in different valence states.

Changes in the valence of a TM ion are usually accompanied by changes in binding energy of its core level photoelectrons. As such, Auger–photoelectron coincidence spectroscopy (APECS) is uniquely well suited to study these materials. In APECS, a core level photoelectron is measured in time coincidence with an Auger electron generated by the decay of the associated core hole. [1–4] By scanning the Auger spectrum in coincidence with photoelectrons of a fixed kinetic energy, the spectrum associated with the decay of a particular core

Correlation Spectroscopy of Surfaces, Thin Films, and Nanostructures. Edited by Jamal Berakdar, Jürgen Kirschner
Copyright © 2004 Wiley-VCH Verlag GmbH & Co. KGaA, Weinheim
ISBN: 3-527-40477-5

hole can be determined. For example, one can obtain an Auger spectrum in coincidence with photoelectrons from a shifted core level of an ion in a reduced oxidation state and obtain a measure of the local electronic structure at that site. In other systems, the magnetic exchange interaction can produce satellite features in the core level spectrum that are associated with different spin couplings between the unpaired core electron and the d-electrons of the valence levels. Thus one may also probe spin-dependent phenomena with APECS.

In this chapter we describe a series of APECS measurements of the maximal valency TM compound rutile TiO_2. With the cation in a formal oxidation state of Ti^{4+}, stoichiometric TiO_2 is expected to be relatively inert chemically. The most interesting interactions are expected to be associated with the occurrence of oxygen vacancies that produce Ti^{3+} ions. In the first part of this study, we have used APECS to determine the line shape of the Ti $M_{23}VV$ Auger spectrum of this system, and find that it has two components associated with two decay channels. In addition, we use the sensitivity of APECS to surface defects to study nucleation sites for sub-monolayer amounts of noble metals (Cu and Ag) deposited on the stoichiometric $TiO_2(110)$ surface. In a previous study we showed that APECS has approximately an order of magnitude more sensitivity to residual surface defects than does conventional photoelectron or Auger electron spectroscopy. In the current work we find that residual surface defects play important, but different roles in determining the initial adsorption behavior of these two systems. Our results provide new insight into recent STM studies of these systems.

13.2 Experimental Aspects

APECS data were acquired at the vacuum ultraviolet (VUV) storage ring of the National Synchrotron Light Source (NSLS) at Brookhaven National Laboratory (BNL). Measurements were performed using two cylindrical mirror analyzers (CMAs) focused at a spot on the sample that was illuminated by monochromatized synchrotron radiation. In the study presented here, the $TiO_2(110)$ surface was excited by 110 eV photons, and Ti 3p core level photoemission spectra were measured in coincidence with Ti $M_{23}VV$ Auger electrons with a kinetic energy of 18 eV. In addition, conventional (singles) Ti 3p photoemission spectra were measured simultaneously. Detailed descriptions of the experimental setup and the basic principles of APECS are presented elsewhere [3–5].

The sample was a commercially produced TiO_2 single crystal whose surface was cut and polished to within $1°$ of the (110) orientation. After insertion into the UHV experimental chamber, a stoichiometric $TiO_2(110)$ surface was prepared by repeated cycles of Ar^+ ion sputtering, followed by annealing to 900 K in vacuum. For the last several minutes of the final anneal cycle, 1×10^{-6} Torr of O_2 was admitted to the chamber while the sample was at 900 K. The sample was then allowed to cool in the presence of the O_2, reaching 400 K in about 60 s, at which time the O_2 was removed from the chamber. The temperature was monitored by a thermocouple attached to a Ta holder in direct contact with the sample. Both resonant valence band and singles core level photoemission spectra indicated that these surfaces were stable for over 24 h. Separate thermal evaporation sources were used to deposit ~ 0.2 ML of Cu or Ag onto the stoichiometric surface. Deposition rates were monitored by a calibrated quartz crystal microbalance, and in all cases, depositions were carried out with the sample at room temperature.

The geometry of the stoichiometric $TiO_2(110)$ surface consists of a planar array of Ti- and O-ions with every other $<001>$ Ti row bonded to bridging oxygens giving alternating rows of exposed 5-fold coordinated and bridge-bonded 6-fold coordinated Ti ions. Both configurations are in the Ti^{4+} formal oxidation state and only a single Ti 3p core level is observed in conventional photoemission. In addition, under certain preparation conditions, quasi-hexagonal "rosette" formations have been observed on the $TiO_2(110)$ surface by STM [6]. Although our sample passed quickly through the optimal conditions for rosette formation, we cannot rule out their presence. Note that the structure preserves the TiO_2 stoichiometry. Furthermore, electronic structure calculations [6] show no signs of electronic states in the band gap, suggesting that the Ti-ions are not reduced. However, this latter point has not been verified experimentally. Upon loss of a bridging oxygen ion, the adjacent Ti-ions are left in the 3+ oxidation state. The spectroscopic signatures of reduced Ti-sites are Ti core levels shifted to lower binding energy and band-gap photoemission resulting from partial occupation of Ti d-levels at chemically reduced sites.

13.3   Results and Discussion

A wide scan singles photoemission spectrum from the stoichiometric surface is shown in Figure 13.1. The identities of the most prominent features are indicated. Note that the Ti $M_{23}VV$ Auger spectrum is just barely visible as a small undulation on a large secondary electron background. As will be seen below, the Ti 3p core level spectrum exhibits almost no change for different surface preparations addressed here. A single peak at ~ 68 eV, corresponding to Ti^{4+} ions, dominates the spectrum. If the surface is intentionally reduced by vacuum annealing, a small but distinct shoulder appears on the high kinetic energy side of the peak. This is primarily from Ti^{3+} ions associated with thermally induced bridging oxygen vacancies. When fractional-monolayer amounts of Cu or Ag are deposited on the surface, very subtle changes in the core level line shape are seen. Ambiguity in these line shape changes makes it extremely difficult to assess how these metals bonds to the surface and, in particular, what bonding, if any, occurs at defect sites.

TiO_2 is formally a maximal valency compound, but there is considerable controversy regarding the presence of a covalent component to the Ti–O bond. Furthermore, $TiO_2(110)$ is the prototypical surface that exhibits photon stimulated desorption of O+ with a threshold corresponding to the Ti 3p core hole excitation via the so-called Knotek–Feibelman mechanism. [7] This process is thought to be mediated by an interatomic Auger decay. Therefore, there is considerable interest in the nature of the valence states on the Ti site, and in the decay processes of the Ti 3p core hole. However, as is clear from Figure 13.1, it is extremely difficult to determine the line shape of the Ti $M_{23}VV$ Auger transition from conventional photoemission or Auger spectroscopy.

In Figure 13.2(a) we display the Ti $M_{23}VV$ Auger spectrum obtained in coincidence 3p photoelectrons from Ti in the 4+ oxidation state. Several important advantages of the APECS measurement are immediately obvious. The coincidence spectrum shows no intensity above 25 eV, illustrating that the portion of the large secondary electron background in Figure 13.1 that is associated with other decay processes in the system is eliminated from the coincidence spectrum. In addition, we see some structure in the spectrum near 20 eV and 15 eV. Finally, the

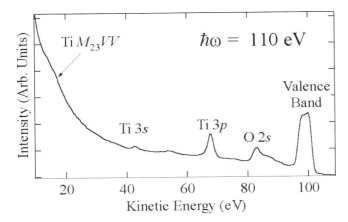

Figure 13.1: A wide scan photoemission spectrum from the $TiO_2(110)$ surface obtained with 110 eV photons. The main spectral features are labeled.

spectrum increases with decreasing kinetic energy below ~ 13 eV. The large intensity in the Auger spectrum at low kinetic energy may be due to secondary electrons created as the Auger electrons travel to and escape from the surface. However, with the high surface sensitivity of APECS it is more likely that this emission is dominated by multiple electron Auger-like decays that leave the system with more than two holes in the final state. If we approximate this background by a simple polynomial and subtract it from the coincidence spectrum, we obtain the intrinsic two-hole Auger shown in Figure 13.2(b). This spectrum has appreciable spectral weight between about 25 eV and 12 eV, and appears to be comprised of two peaks, one centered at ~ 19 eV, and the other at ~ 15 eV. Even without attempting to model the line shape of this spectrum, the coincidence Auger spectrum exhibits an unexpected feature. If one considers the binding energy of the Ti 3p core level and the location of the Fermi level of the system, the predicted threshold for the Ti $M_{23}VV$ Auger decay is 27 eV. The observed threshold, however, is 25 eV, shifted a full 2 eV from what is expected from simple conservation of energy.

Modeling the Auger spectrum reveals more surprises. If the $M_{23}VV$ Auger decay were dominated by an interatomic Auger decay, the electron that fills the Ti core hole, and the Auger electron that is ejected from the system, would come from the O 2p levels. These orbitals are relatively extended and thus correlation effects are expected to be small. In such a case, the line shape of the Auger spectrum is expected to be the self-convolution of the O 2p band of TiO_2. The predicted Auger spectrum based on this model is shown as the dotted curve in Figure 13.2(b). Although this curve clearly overlaps the experimental data, there are still significant problems. First, the threshold is at the predicted value, which is at a higher kinetic energy than the threshold of the data. Secondly, the leading edge of the calculated curve occurs at a significantly higher energy than does the experimental data and, in fact, the model has a peak and begins to decrease before the data have even reached their first maximum. In addition, the calculated curve goes to zero near 16 eV and does not account for the experimental intensity that persists until $\sim$12 eV.

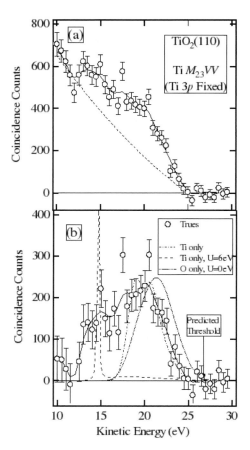

Figure 13.2: (a) The Ti $M_{23}VV$ Auger spectrum from the TiO$_2$(110) surface obtained in coincidence with Ti 3p photoelectrons. The dashed curve is an estimate of the background associated with Auger-like transitions involving more than two electrons. The solid curve is a smooth to the as-acquired spectrum. (b) The background subtracted Ti $M_{23}VV$ coincidence Auger spectrum. The expected threshold for the spectrum, based on the core level binding energy and location of the Fermi level, is indicated. Model calculations of the Auger line shape, based on different assumptions (see text), are also shown.

A second approach is to consider what is expected if there is a significant covalent component to the Ti–O bond. One can then consider the occupied valence density of states on the Ti site and then apply the self-convolution. The results of such a calculation, using the local density of states from the literature, [8–11] are shown as the dot-dashed curve in Figure 13.2(b). Clearly, this notion resolves many of the problems encountered by the previous approach. The curve exhibits a threshold, a rising edge, and a peak energy that all agree well with the experimental data. This model gives a good account for the experimental data above ~ 16 eV, but still does not account for considerable experimental intensity at lower energies. It is wellknown, however, that the presence of a significant Coulomb interaction between the final state core holes in a *CVV* Auger transition will shift spectral weight to lower kinetic energy. Furthermore, if the coulomb energy, U, is larger that the band width, W, a bound state is split off below the lower edge of the band. A typical Coulomb interaction energy for a transition metal is ~ 6 eV. If we apply the wellknown Cini–Sawatzky transformation to the Ti SCDOS, the long dashed curve in Figure 13.2(b) is produced. For this value of U, the split off state occurs at ~ 15 eV and appears to account nicely for the extra emission that is not described by the SCDOS alone. Clearly, the calculated curve is much narrower than the experimental one, but this may be related to the fact that this split-off state is degenerate with other decay channels (such as the multiple-electron Auger decay channels discussed above) to which it may couple, shortening its lifetime and giving it a larger width. The combination of these two decay channels, one given by the SCDOS on the Ti site, and the other described by the Cini–Sawatzky transformed line shape, appears to account for the experimental data. The different correlation energies of the two decay channels suggest that there are two distinct ways in which the two holes in the Auger final state can evolve. The typical Ti-ion is bound to six oxygen nearest neighbors. Owing to the covalent component in the O–Ti bond, a hole in the Ti can hop to a neighboring O-ion. Then, the second hole has a one-in-six chance of following the first hole onto the same O-site, or it may migrate to any of the other five oxygen neighbors. In the former case, the holes remain together and the Auger spectral weight is shifted to lower kinetic energy. In the latter case, the holes propagate independently, resulting in the more band-like portion of the spectrum.

The coincidence Auger spectra discussed above show how APECS can eliminate the uncorrelated secondary electron background, making it much easier to determine the intrinsic line shape of the Auger transition. We have found in a previous APECS study of $TiO_2(110)$, however, that so-called "reverse scans", that is coincidence photoemission spectra, are particularly sensitive to reduced Ti sites on this surface [12]. Owing to the fact that the valence electron density is very low ($\sim 0.1 \ e$) at Ti^{4+} sites, the probability for Auger decay is relatively small. In contrast, when one considers a reduced site, as is typically found adjacent to a defect (which is typically an oxygen vacancy), the electron density is much higher ($\sim 1 \ e$) and the Auger decay rate can be enhanced by one to two orders of magnitude. As a result, reverse scans on this system (i.e., coincidence photoelectron spectra) are particularly sensitive to surface defects in this system.

Coincidence Ti 3p core level photoemission spectra are shown in Figure 13.3. Figure 13.3(a) shows the spectrum obtained from the clean, stoichiometric $TiO_2(110)$ surface [12]. Although in the singles Ti 3p spectrum we find $< 2\%$ contribution from reduced Ti-ions, the coincidence spectrum has a substantial amount of such intensity. In a previous study of this surface [5], we fit the spectrum with three Gaussians centered at the energies corresponding to the $4+$, $3+$, and $2+$ oxidation states of Ti. The fit indicates that $\sim 10 \pm 2\%$ each of the total coincidence Ti 3p intensity comes from Ti^{3+} and Ti^{2+} oxidation states. We associate this emission with residual (or intrinsic) defects that are present on the stoichiometric surface. Our previous work indicates that these sites are either from step edges, from isolated bridging oxygen vacancies on the terraces, or from strands that contain Ti_2O_3-like configurations that protrude from the rosette structures on the surface [6, 12–16].

The coincidence Ti 3p spectrum changes significantly after deposition of either Cu or Ag. Figure 13.3(b) shows the Ti 3p coincidence spectrum obtained from the 0.2 ML $Cu/TiO_2(110)$ surface. The fit to the spectrum indicates that the percent of spectral weight associated with Ti^{4+} and Ti^{3+} oxidation states is about 80% and 20%, respectively, while the Ti^{2+} intensity is below detection limits ($< \sim 3\%$). This means that either the concentration of exposed Ti^{3+} is significantly enhanced, or that the concentrations of exposed Ti^{4+} and Ti^{2+} sites are reduced. The creation of a substantial concentration of exposed Ti^{3+} is unlikely for two reasons. First, based on our previous studies, the extent of reduction necessary to produce such a change in the coincidence spectrum would be sufficient to produce a significant Ti^{3+} feature in the Ti 3p singles spectrum. Instead, the line shape of the singles core level spectrum for this

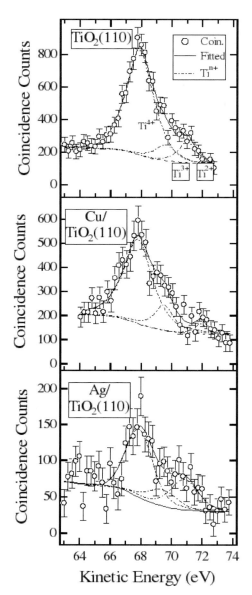

Figure 13.3: (a) The Ti 3p photoelectron spectrum from the stoichiometric $TiO_2(110)$ surface obtained in coincidence with Ti $M_{23}VV$ Auger electrons. The contribution from various Ti oxidation states is indicated. (b) Same spectrum as (a) but after deposition of 0.2 ML Cu. (c) Same spectrum as (a) but after deposition of 0.2 ML Ag

surface is almost identical to that of the clean TiO_2 surface. Secondly, resonance valence band photoemission spectra from the $Cu/TiO_2(110)$ surface (not shown) exhibit no significant emission above the valence band maximum. From this we conclude that Ti^{2+} ions act as nucleation sites for Cu cluster formation. As mentioned above, Ti^{2+} is not expected along well-formed steps or rosette structures. This suggests that Cu primarily bonds at non-ideal step sites such as kinks and reduced portions of the rosettes, and is unlikely to bond at isolated bridging oxygen vacancies. As a result, one expects a low concentration of Cu islands on perfect regions of the terrace. The associated reduction of Ti^{4+} sites suggests that the Cu islands, regardless of their shape, obscure a fraction of the five-fold Ti^{4+} sites. We also note that the suppression of the Ti^{2+} emission enables us to see the contribution from Ti^{1+} in the coincidence spectrum.

The initial growth of Ag on the $TiO_2(110)$ surface provides an interesting contrast to that of Cu. Figure 13.3(c) shows the Ti 3p coincidence spectrum obtained after the deposition of 0.2 ML of Ag on the $TiO_2(110)$ surface. Once again, even before a quantitative assessment, it is clear that the coincidence Ti 3p line shape for this system is distinct from that of the clean surface and from that of the Cu-adsorbed system. Fitting this spectrum shows that the fraction of spectral weight associated with Ti^{2+} has approximately doubled from its value for the clean surface. From this we can conclude that Ag prefers to bond at Ti^{3+} sites, which are now obscured and thus do not contribute to the spectrum, in contrast with Cu which preferred Ti^{2+}. Our understanding of the spectrum of Figure 13.3(a) [12] leads to the conclusion that Ag will have a higher propensity than Cu to nucleate at residual bridging oxygen vacancies on the terrace, at Ti^{3+} defects as steps, or at Ti^{3+}-sites associated with the Ti_2O_3-like strands that extend from the rosette structures. Similar to the results for Cu-adsorption, Ti^{4+} intensity is suppressed at a rate more rapid than the reduced Ti sites, suggesting that, after nucleation at the reduced site, an Ag island begins to extend over the terrace.

Our APECS shed some interesting light on recent STM studies of both the $Cu/TiO_2(110)$ and $Ag/TiO_2(110)$. [17, 18] At low coverages, both Ag and Cu clusters tend to nucleate at step edges. However, for a given coverage, Cu tends to form fewer and larger clusters, while the Ag clusters are smaller and more evenly distributed along a step. This is consistent with Cu bonding to the less common Ti^{2+} site and Ag to the more common Ti^{3+} site. At higher coverages, Ag appears to find the occasional Ti^{3+} site on the terrace, while Cu is much more likely to remain at the steps, or at the more textured rosette structures, where it presumably is more likely to find Ti in the lower Ti^{2+} oxidation state.

13.4   Conclusions

We have measured the Ti $M_{23}VV$ Auger spectrum of the $TiO_2(110)$ surface in coincidence with Ti 3p photoelectrons. We find that the spectrum has two components, one associated with a decay where the two core holes propagate together and the other where they separate onto different O-sites. We have also measured the Ti 3p photoemission spectrum in coincidence with Ti $M_{23}VV$ Auger electrons for both the clean $TiO_2(110)$ surface and the surface covered with 0.2 ML of Ag or Cu. We find that Ag clusters tend to nucleate at sites in a Ti^{3+} oxidation state while Cu nucleates at Ti^{2+} sites. These results help explain recent STM data from these metal-covered surfaces.

Acknowledgment

Support from grants NSF-DMR98-01680 and the Petroleum Research Fund under grants PRF$^{\#}$ 33750-AC5,6 and PRF$^{\#}$ 40236-AC5S is gratefully acknowledged. The NSLS at BNL is supported by the U.S. DoE under Contract No. DE-AC02-98CH10886.

References

[1] H. W. Haak, G. A. Sawatzky, and T. D. Thomas, Phys. Rev. Lett. **41**, (1978) 1825.

[2] H. W. Haak, G. A. Sawatzky, L. Ungier, K. Ginzewski, and T. D. Thomas, Rev. Sci. Instrum. **55**, (1984) 696.

[3] E. Jensen, R. A. Bartynski, S. L. Hulbert, and E. D. Johnson, Rev. Sci. Instrum. **63**, (1992) 3013.

[4] R. A. Bartynski, E. Jensen, and S. L. Hulbert, Phys. Scr. **T41**, (1992) 168.

[5] A. K. See, W.-K. Siu, R. A. Bartynski, A. Nangia, A. H. Weiss, S. H. Hulbert, X. Wu, and C.-C. Kao, Surf. Sci. Lett. **383**, (1997) L735.

[6] M. Li, W. Hebenstreit, L. Gross, U. Diebold, M. A. Henderson, D. A. Jennison, P. A. Schultz, and M. P. Sears, Surf. Sci. **437**, (1999) 173.

[7] M. L. Knotek and P. J. Feibelman, Phys. Rev. Lett. **40**, (1978) 964.

[8] S. Munnix and M. Schmeits, Phys. Rev. B **28**, (1983) 7342.

[9] S. Munnix and M. Schmeits, Phys. Rev. B **30**, (1984) 2202.

[10] S. Munnix and M. Schmeits, Phys. Rev. B **31**, (1985) 3369.

[11] S. Munnix and M. Schmeits, J. Vac. Sci. Technol. A **5**, (1987) 910.

[12] W. K. Siu, R. A. Bartynski, and S. L. Hulbert, J. Chem. Phys. **113**, (2000) 10697.

[13] U. Diebold, J. F. Anderson, K.-O. Ng, and D. Vanderbilt, Phys. Rev. Lett. **77**, (1996) 1322.

[14] H. Onishi and Y. Iwasawa, Phys. Rev. Lett. **76**, (1996) 791.

[15] M. Li, W. Hebenstreit, and U. Diebold, Surf. Sci. Let. **414**, (1998) L951.

[16] M. Li, W. Hebenstreit, and U. Diebold, Phys. Rev. B **61**, (2000) 4926.

[17] D. Chen, M. C. Bartlet, R. Q. Hwang, and K. F. McCarty, Surf. Sci. **450**, (2000) 78.

[18] D. A. Chen, M. C. Bartlet, S. M. Seutter, and K. F. McCarty, Surf. Sci. **464**, (2000) L708.

14 Relevance of the Core Hole Alignment to Auger–Photoelectron Pair Angular Distributions in Solids

Giovanni Stefani, Roberto Gotter, Alessandro Ruocco, Francesco Offi, Fabiana Da Pieve, Alberto Verdini, Andrea Liscio, Stefano Iacobucci, Hua Yao, and Robert Bartynski

Auger–photoelectron coincidence spectroscopy (APECS) has shown that singling out Auger spectra individual components on the basis of the related photoelectron, i.e., defining the parent core ionic state, increases the energy discrimination capability of both Auger and photoelectron spectroscopy. Angle resolved APECS (AR-APECS) on Ge(100) surface has shown peculiar new features of the electron pair angular distribution with Auger detection angle. It was speculated that these features were to be ascribed to the alignment of the core-hole state induced by selection of the angle of the photoelectron. Recent results on Ge(100) and Cu(111) have drawn attention to the possibility of selecting specific magnetic sublevels of both Auger and photoelectron wavefunctions. Good use of this capability has been made in singling out otherwise overlapping components of both Ge L_3VV and Cu M_3VV multiplet split spectra, thus demonstrating that resolution in angle improves discrimination in the energy scale as well.

14.1 Introduction

To develop nanoscale aggregates often requires characterizations with discrimination, both in energy and space, that are beyond the capabilities of well-established spectroscopies. This is the main motivation for the progress in the fine analysis of matter that has occured over the past decade. Spatial, energy and time resolution have been improved by exploiting the unprecedented brightness of third generation synchrotron radiation sources. Among these achievements, Auger–photoelectron coincidence spectroscopy (APECS) has its own individual status as it accesses information that is not merely a refinement of what individual Auger and photoemission spectroscopy would access. An APECS experiment detects, correlated in time and selected in energy, electron pairs generated by the creation–relaxation process of a core hole in a specific atomic site of a solid. APECS combines in a unique way the Auger (AES) sensitivity to valence (shallow core) density of states with the atomic site selectivity of core photoemission (XPS) and has been shown to gain considerable discrimination in probing electron correlation and charge transfer phenomena on the femtosecond time scale. Coster–Kronig transitions are the natural candidate for experiments aimed at studying electron correlation but not even APECS is capable of distinguishing individual components of the lifetime broadened multiplet structure generated by their two holes final states. In this chapter we propose to use the angular correlation between the two electrons (Auger and photoelec-

Correlation Spectroscopy of Surfaces, Thin Films, and Nanostructures. Edited by Jamal Berakdar, Jürgen Kirschner
Copyright © 2004 Wiley-VCH Verlag GmbH & Co. KGaA, Weinheim
ISBN: 3-527-40477-5

tron) in order to gain further discrimination in disentangling individual multiplet components that are separated in energy by less than their lifetime broadening. This idea steams from the observation that the amount of information obtained from photoemission spectra is vastly increased when one measures the angular distribution of photoemitted electrons as well as their energy. For example, ultraviolet photoemission (UPS) gives information about the valence-band density of states of a solid, while angle resolved UPS enables one to directly map the energy bands. Similarly, from XPS and AES one can obtain information about the chemical state of a surface, while angle-resolved measurements (i.e. photoelectron (PED) and Auger electron diffraction (AED)) enable one to perform surface structural measurements and produce holographic images of surface geometry. In a similar way AR-APECS is expected to add an important level of discrimination to the APECS technique. This capability has been extensively applied to investigate atoms and molecules in the gas phase, and the extra discrimination originated by resolving in angle, the electron pair probability distribution, has given access to fine details of the core ionization dynamics such as core state polarization, post collisional electron–electron interaction, and quantum interference between continuum final states [1]. The main goal of the present chapter is to show that a similar level of discrimination can also be achieved in solids. This will be done by discussing recent AR-APECS results on Ge(100) and Cu(111) surfaces. To perform an AR-APECS experiment amounts at measuring the multiple differential cross section dI_{Au}, angular ($d\Omega_{Au}$ and $d\Omega_{ph}$) and energy (dE_{ph}) resolved, as

$$\frac{dI_{Au}}{d\Omega_{Au}d\Omega_{ph}dE_{Au}} \cong \Gamma(E_{Au}, E_{Au}^0)|T_{\beta\alpha}|^2 \tag{14.1}$$

where $\Gamma(E_{Au}, E_{Au}^0)$ is the Auger lineshape and $T_{\beta\alpha}$ is the Auger decay amplitude. It is therefore evident that several different modes are available for such an experiment depending upon the various parameters that are kept fixed while varying the others. Here we shall report on two different modes:

(i) *Energy mode*: by fixing the direction of detection for the two electrons and by scanning the Auger energy a "selected" Auger line shape is measured.

(ii) *Angular mode*: by fixing the Auger energy, the direction of detection of the photoelectron and by scanning the Auger ejection angle a "selected" AED is measured.

Both modes will amount to select and control core hole state magnetic sublevels occupancy, hence inducing an apparent variable alignment of the intermediate state of the process. It will be shown that this selectivity enables discrimination of multiplet components overlapping in energy. Finally, the extreme surface sensitivity of an AR-APECS experiment will be demonstrated with the help of measurements done on the Cu(111) surface.

14.2   AR-APECS Two Step Model

To interpret the main features of AR-APECS in solids, we propose to use a two-step model. The first step describes the core ionization process and the following Auger decay as if it were happening in an isolated atom. In this step, the energy, the angular moments and magnetic quantum numbers that describe the wavefunctions of the two outgoing electrons are defined. These atomic wavefunctions will act as source waves for the following interaction with the crystal lattice. The interaction is assumed to be elastic, hence well described by a diffraction process that ultimately will determine the flux of probability reaching, at infinity, the two electron analyzers used to select (in angle and energy) and detect the Auger–Photoelectron pair distributions. In the first atomic step, the core ionization process will be described within the dipole approximation and for linearly polarized light. In calculating the Auger decay probability and the angular correlation between the two final electrons, the approach proposed in Ref. [2] will be adopted and extended to open shell atoms. This straightforward extension is of particular relevance in the solid matter, where open shell systems are mostly studied, and the closed shell description, so far used for isolated atoms, will be inadequate.

14.2.1 Atomic Core Ionization and Relaxation

At the atomic site where core ionization takes place, the angular distribution of the electron pair describing the correlation source wavefunction can be obtained using spherical tensors theory [3]. The anisotropy of the initial emission is determined by the angular momentum components which describe the outgoing electrons and by the radial matrix elements which govern the process.

Let us consider the photoionization and decay as independent processes:

$$
\begin{aligned}
h\nu(j_\gamma) + A(J_0) &\longrightarrow A^+(\alpha_1 J_1, \alpha_1' J_1') + e_1^-(j_1) \\
&\hookrightarrow A^{++}(\alpha_2 J_2, \alpha_2' J_2') + e_2^-(j_2)
\end{aligned}
\tag{14.2}
$$

where J_0, j_γ are the angular momenta related to the initial atom and to the photon, $\alpha_1 J_1$, $\alpha_1' J_1'$ and $\alpha_2 J_2, \alpha_2' J_2'$ are related to the ions A^+, A^{++} that can be created in different angular states, and j_1, j_2 are the angular momenta of the photoelectron and Auger electron.

Using the dipole approximation and in the reference frame in which the polarization vector is parallel to the z axis, the statistical tensors [2] of the photoionized state A^+ that describe the distribution of the subsequent Auger electrons can be written as:

$$
\rho_{k_1 q_1}(\alpha_1 J_1, \alpha_1' J_1') = \sum_{k_e q_e k_r q_r} \frac{1}{2J_0 + 1} \sqrt{\frac{4\pi}{2k_e + 1}} Y_{k_e q_e}(\theta_e, \phi_e)
\tag{14.3}
$$
$$
\times (k_1 q_1, k_e q_e | k_r q_r) \rho_{k_\gamma q_\gamma} \mathcal{I}_{k_e k_1 k_r}
$$

where $\rho_{k_\gamma q_\gamma}$ are the statistical tensors related to the photon, $Y_{k_e q_e}(\theta_e, \phi_e)$ are the spherical harmonics of the photoelectron, and $(k_1 q_1, k_e q_e | k_r q_r)$ are the Clebsch–Gordan coefficients. The dynamical parameters $\mathcal{I}_{k_e k_1 k_r}$ are related to the dipole matrix elements through the relation:

$$
\mathcal{I}_{k_e k_1 k_r} = \frac{1}{4\pi} \hat{k}_1 \hat{k}_e \sum_{l_1 l_1' j_1 j_1' J_r J_r'} (-1)^{J_r + J_0 + k_r + j_1' - \frac{1}{2}} \hat{J}_r \hat{J}_r' \hat{l}_1 \hat{l}_1' \hat{j}_1 \hat{j}_1' (l_1 0 l_1' 0 | k_e 0)
$$
$$
\times \begin{Bmatrix} j_1 & l_1 & \frac{1}{2} \\ l_1' & j_1' & k_e \end{Bmatrix} \begin{Bmatrix} 1 & J_r & J_0 \\ J_r' & 1 & k_r \end{Bmatrix} \begin{Bmatrix} J_1 & j_1 & J_r \\ J_1' & j_1' & J_r' \\ k_1 & k_e & k_r \end{Bmatrix}
\tag{14.4}
$$
$$
\times D_{J_1 j_1 J_r J_0} D^*_{J_1' j_1' J_r' J_0}
$$

where the notation used for dipole matrix elements $D_{J_1 j_1 J_r J_0}$ implies that the angular momentum J_1 is coupled to the photoelectron momentum j_1 through a resonant state J_r, which in turn is determined by coupling J_0 with j_γ. The angular correlation function between Auger and photoelectrons can be obtained by combining these statistical tensors with the tensors [4] that describe spin-unresolved detection efficiency:

$$W = \sum_{k_1 q_1 J_1 J_1'} \frac{\hat{J_1}^{-1}}{4\pi} \rho_{k_1 q_1}(\alpha_1 J_1 \alpha_1' J_1') R_{k_1}(J_1 J_1'; J_2) \sqrt{\frac{4\pi}{2k_1+1}} Y_{k_1 q_1}(\theta_2, \phi_2) \quad (14.5)$$

where $Y_{k_1 q_1}(\theta_2, \phi_2)$ are the spherical harmonic of the Auger electron and the *Auger parameter* $R_{k_1}(J_1, J_1'; J_2)$ is related to the Auger matrix elements $V_{J_2 j_2 J_1}$ as follows:

$$R_{k_1}(J_1 J_1'; J_2) = \sum_{l_2 l_2' j_2 j_2'} (-1)^{J_1 + J_2 + K_1 - \frac{1}{2}} \hat{l_2} \hat{l_2'} \hat{j_2} \hat{j_2'} \hat{J_1} (l_2 0 l_2' 0 | k_1 0)$$

$$\times \left\{ \begin{array}{ccc} J_1 & j_2 & J_2 \\ j_2' & J_1' & k_1 \end{array} \right\} \left\{ \begin{array}{ccc} l_2 & j_2 & \frac{1}{2} \\ j_2' & l_2' & k_1 \end{array} \right\} V_{J_2 j_2 J_1} V_{J_2 j_2' J_1'}^* \quad (14.6)$$

The angular correlation function (14.5) describes the anisotropy of the initial emission at the atomic site and takes into account, for both electrons, all the possible angular momenta permitted by the selection rules. The resulting distribution of the correlated electrons represents the source wave that will be diffracted from the periodic potential of the crystal. Relation (14.5) shows that different angular components contributes in a different way to the intensity measured at given detection angles. We also note that the weight of the different angular components is largely determined by sums of Clebsch–Gordan coefficients, that explicitly depend on the magnetic quantum number m_l. Hence, selecting specific ejection direction for the electron pairs amounts to selecting dominant l and m_l values for the wavefunctions that describe, separately, the Auger–electron and the photoelectron.

14.2.2   Diffraction from Crystal Lattice

The features of the source wavefunctions are relevant for the scattering process since the potential describing the interaction with the atoms of the crystal is determined by an attractive Coulomb part and a repulsive barrier which depends on the orbital momentum of the electrons. For electrons with high orbital momentum, the potential has a non-negligible repulsive barrier outside the attractive well. Simulations of the diffraction pattern show that the angular distributions depend on both orbital and magnetic quantum numbers l and m_l [5]. The m_l dependence can be generated by a non statistical distribution of initial sublevels (as happens in magnetic materials) or by excitation through light of specific polarization (which permits emission from certain sublevels). In both cases, the features of the final state reflect the different non-statistical weights of the partial waves. Through the introduction of a quantization axis, we can obtain the non-statistical contributions of the different sublevels, even in non-magnetic materials [6].

In our case, where we are interested in the emission of electron pairs, we could select a particular angular component of the wave related to the first electron by revealing it at a certain

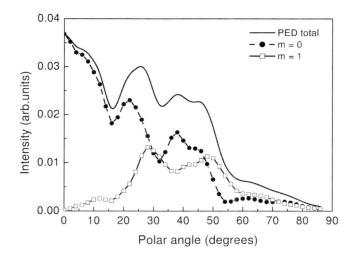

Figure 14.1: Different contributions of m_l distributions to the total intensity in the second order of scattering; simulation is made with the MSCD code on direction $(-1, -1)$ of the surface Cu(111) with a photoelectron energy $E = 80$ eV.

angle. This selection determines the partial wave components of the Auger electron wavefunction since the possible contributions are related to those of the photoelectron by the selection rules. For linearly polarized light, the interdependence of l and m_l for the two electrons is expressed in Eq. (14.7), where $l_1, m_1, l_2, m_2, l_c, m_c, l_{e_2}, m_{e_2}$ are the orbital momentum of the two final holes, of the hole left behind by core ionization and of the Auger electron respectively. In this way the diffraction pattern could be interpreted, in the two step model, as generated by the diffraction of the two source waves whose l and m_l quantum numbers are determined by

$$l_c - |l_1 - l_2| \leq l_{e_2} \leq l_c + l_1 + l_2$$
$$l_c + l_1 + l_2 + l_{e_2} = \text{ even} \tag{14.7}$$
$$m_c + m_1 = m_2 + m_{e_2}$$

To elucidate the relevance of selecting l and m_l quantum numbers, we present a simulation of the diffraction patterns of a photoelectron originating from the 3p orbital in the Cu(111) surface, with a kinetic energy of 80 eV. The simulation has been performed with MSCD (multiple scattering calculation of diffraction) program code [7] that simulates the elemental and state-specific core-level photoelectron diffraction pattern from a solid; it is based on the Rehr–Albers separable representation of spherical-wave propagators [8]. The calculation was performed using a muffin-tin potential, the dipole matrix elements have been derived in frozen core approximation from the Hermann–Skillmann tables. The l and m_l values of the source wavefunctions were defined by dipole selection rules and by (14.7); the results are reported in Figure 14.1. The full curve displays the calculated photoemitted intensity as a function of the polar angle in the $(-1, -1)$ direction. The MSCD code allows one to evaluate the contribution of the individual m_{e_1} components of the source wavefunction (with $m_{e_1} = m_c$ in the case of linearly polarized light) to the total angular distribution probability, that are shown by dashed curves in the figure. From the figure it is evident that different l, m_l partial source waves contribute with different weight to the probability of emitting the photoelectron at a given angle. Similar results are obtained by applying the MSCD code to calculate the Auger electron diffraction pattern. Hence, to detect a pair of photo–Auger electrons at specific azimuthal and polar angle with respect to light polarization and surface orientation, amounts to highlighting the contribution of a specific subset of partial source waves among those which are allowed by selection rules (14.7). Moreover, always through relations (14.7), it is evident that to select l and m_l components for the final electrons results in selecting l_c, m_c, i.e. different magnetic sublevels of the core hole states. In other words, we can expect that even in the solid state the polarization of the ion state created by the photoionization process reflects itself in the emission of the electrons in preferential directions. From this observation comes an interest in studying how the selectivity in the core hole state sublevels, that is characteristic of an AR-APECS experiment, will reflect itself in the angular and energy distribution of the electron pairs. This subject will be addressed in the following discussion of the results of two AR-APECS experiments on Ge(100) and Cu(111) performed in angular and energy mode respectively.

14.3   Experimental Results

Measurements were performed using the unique capabilities of the ALOISA beamline at the ELETTRA synchrotron in Trieste, Italy. The experimental setup is discussed in detail elsewhere [10] and only a brief description will be given here. The monochromatic, linearly polarized radiation impinges on the sample at grazing incidence (i.e. below the critical angle) and the surface normal lies in the plane determined by the photon beam and its polarization vector (nearly p-polarization for experiments herewith reported). Seven electron analyzers are devoted to simultaneously acquire ten different coincident pairs. Two of them (termed bimodal) rotate around the photon beam axis and around an axis normal to it (see Figure 14.2). These two analyzers are usually employed to perform a polar scan in the scattering plane defined by the surface normal and the photon beam axis. The other five analyzers (18° apart and termed axial) are positioned on a plane containing the photon beam axis that can rotate around it. In this way the two bimodal analyzers measure an angular distribution in coincidence with five different values of the momentum wavevector selected by the five axial analyzers. The experimental data were acquired in two modes: an integrated mode where an Auger electron was detected by one of the bimodal analyzers and a photoelectron was detected in any of the five axial analyzers, and a pairwise mode where an Auger electron detected in one of the bimodal analyzers comes in coincidence with photoelectrons in only one particular axial analyzer. In both modes, timed spectra for each pair of analyzers, covering a range of several hundred nanoseconds on either side of $\Delta t = 0$, were recorded so that the accidental contribution to the coincidence signal could be determined and subtracted to produce the true coincidence signal, which is reported here. We simultaneously record a non-coincidence, or singles, energy or angle Auger pattern during the AR-APECS measurement.

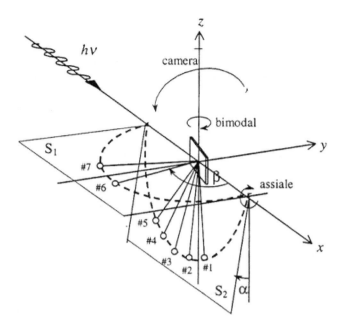

Figure 14.2: Schematics of the rotating frames of analyzers of the ALOISA multicoincidence end station. The S1 plane (bimodal plane) contains two analyzers that are moved by combining band g rotations. The S2 plane (axial plane) rotates around the photon beam axis and contains five analyzers. If the bimodal analyzers detect the Auger electrons, the axial analyzers detect the photoelectrons. Auger electrons detected by each individual bimodal (axial) analyzer are correlated in time with photoelectrons detected by whichever axial (bimodal) analyzer.

14.3.1 Angular Discrimination

In the previous section we already outlined the foundations on which the interpretation of the experiments reported here rest. Individual photoelectrons and Auger electrons are emitted with an angle distribution that results from sampling all possible final states that conserve energy and momentum. Therefore, in a conventional non-coincidence experiment, the intensity of Auger (photoelectron) current into a given solid angle is the result of a suitable averaging over all energies and angles allowed to the partner photoelectron (Auger). In a coincidence experiment, only the subset of events for which both electrons (Auger and photoelectron) fall within the angle and energy windows accepted by the analyzers will contribute to the measured probability current of detecting time correlated pairs. Let us start by discussing the correlated behavior of the $L_3 M_{45} M_{45}$ Auger angular distribution measured in coincidence with $2p_{3/2}$ photoelectrons from the Ge(100) surface when excited by a 1450 eV linearly polarized photon beam [11]. The experiment was performed in a pair-wise angular mode (PAM) and the results are reported in Figure 14.3. The data points with error bars are the coincidence distributions and the dashed line is a guide to the eye. The main differences with respect to the AED, reported at the bottom of the figure, and with respect to each other are sharp and

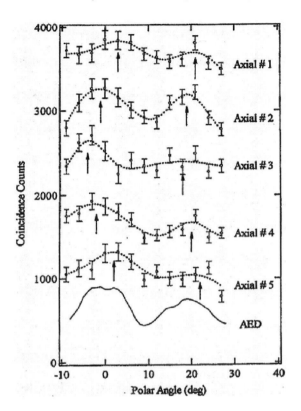

Figure 14.3: Data points are the Pairwise Angular Mode AR-APECS angular distribution of Ge $L_3M_{4,5}M_{4,5}$ Auger electrons measured in coincidence with Ge $2p_{3/2}$ core photoelectrons detected by each individual axial analyzer. Error bars are statistical uncertainties. The noncoincidence Auger angular distribution AED was measured at the same time and by the same bimodal analyzers used to collect the AR-APECS signal.

beyond statistical uncertainty. The differences between these coincidence distributions collected in PAM can be understood using the two-step model described in the previous section. This approach is justified by the long lifetime of the core hole and the well-defined parity and angular momentum of the final state [12]. In this experiment, we are detecting a 2p photoelectron, so from dipole approximation we infer that $l_c = 1$. Furthermore, the $L_3M_{45}M_{45}$ Auger transition leaves two 3d holes (i.e. $l_1 = l_2 = 2$) in the final state and from relations (14.7) l_{e_2} may assume the value of 1, 3, or 5. AED studies of similar transitions indicate that the Auger electron is predominantly $l_{e_2} = 3$. This limitation, together with the conditions imposed by (14.7), enable us to specify that

$$m_{e_2} = m_1 - m_2 \qquad \text{if} \qquad m_c = 0 \tag{14.8}$$

$$m_{e_2} = m_1 - m_2 + 1 \qquad \text{if} \qquad m_c = 1 \tag{14.9}$$

$$m_{e_2} = m_1 - m_2 - 1 \qquad \text{if} \qquad m_c = -1 \tag{14.10}$$

In the coincidence measurement we detect only Auger electrons associated with the $^1G_4$ configuration of the two hole final state. Therefore, the Auger final state has $L = 4$ and $m_L = 0, \pm 1, \pm 2, \pm 3$, and ± 4. These sublevels are given by the particular combinations of $|l_1 m_1\rangle, |l_2 m_2\rangle$ product states specified by the appropriate Clebsch–Gordan coefficients. We can therefore determine the relative probability for each possible value of m_{e_2} for a given value of m_c. In the coincidence measurements here reported, the five axial analyzers, tuned on the $2p_{3/2}$ photoline, detected events at different emission angles, and therefore will have different relative weightings of $m_c = 0$ or ± 1. As discussed in the previous paragraph, Auger electrons with different values of m_{e_2} have different angular distributions; hence we expect that each PAM angular distribution will have a different profile. In effect, regarding the maximum near $\Theta = 0°$, the experiment reported in Figure 14.3 shows that for axial analyzers 1 and 5 this feature is on the positive-angle side of the surface normal while for analyzer 3 is on the negative side. The curves from analyzers 2 and 4 are intermediate in this respect. Although less pronounced, the feature near $\Theta = 20°$ appears to exhibit a similar trend. PAM investigations on Ag(4p) and Si(2p) gave rise to similar observations of sharp differences between coincidence and non-coincidence Auger angular distributions. This was not the case for C(1s) where no l and m_l selectivity is introduced by the time correlated detection of the two final electrons. All these findings are well accounted for by the proposed two-step model and support the hypothesis that in solids and atoms alike, by measuring coincidence angular distributions different "alignment" for the intermediate core hole state can be selected.

14.3.2 Energy Discrimination

LVV and MVV Auger transitions in solid Cu involve two final holes in the valence band. However, since the d band is quite narrow, the holes remain localized, and the spectrum can be interpreted in terms of angular momentum coupled two particle atomic states taking into account all the associate multiplet terms transition. In the following we will focus our attention on the transition $M_3 M_{4,5} M_{4,5}$ in Cu(111): the possible partial waves for the photoelectron are $l_{e_1} = 0, 2$ while for the Auger electron we obtain $l_{e_2} = 1, 3, 5$. The azimuthal quantum numbers of the two electrons are related by relations (14.7).

The possible final multiplet terms related to the ion A^{++} in the configuration $3d^8 4s$ are: $^1S_0(l_{e_2} = p_{1/2,3/2}), ^1G_4(l_{e_2} = f_{5/2,7/2}h_{9/2,11/2}), ^1D_2(l_{e_2} = p_{1/2,3/2}), ^3P_{0,1,2}(l_{e_2} = p_{1/2,3/2}), ^3F_{2,3,4}(l_{e_2} = f_{5/2,7/2})$ where in parenthesis we have indicated the possible partial waves for the Auger electron. Theoretical calculations and experimental evidence show that the dominant multiplet terms are $^1G_4(l_{e_2} = f_{5/2,7/2}h_{9/2,11/2})$ and $^3F_{2,3,4}(l_{e_2} = f_{5/2,7/2})$ [13, 14] and that the dominant partial wave for the Auger electron is consequently $l_{e_2} = 3$. In the following we will consider only this two multiplet terms. In LS coupling, which is appropriate for these transitions [13], each multiplet term can be written in terms of the states $|l_1 m_1\rangle |l_2 m_2\rangle$ of the two holes through the coupling coefficients:

$$|LM\rangle = \sum_{l_1+l_2=L, m_1+m_2=M} C^{LM}_{l_1 m_1 l_2 m_2} |l_1 m_1\rangle |l_2 m_2\rangle. \qquad (14.11)$$

Taking into account the linear polarization of the light and the l, m_l selectivity expected from AR-APECS, it is foreseen that the relative contribution of the ^1G and ^3F to the coincidence Auger line shape changes when changing the angles at which Auger and photoelectrons

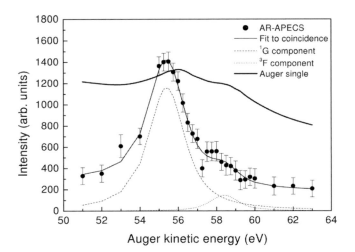

Figure 14.4: Data points are the Cu(111) AR-APECS spectrum measured in integrated energy mode; i.e. the energy distribution of the M_3VV Auger electron (measured by bimodal analyzers) in coincidence with a $3p_{3/2}$ photoelectron (see text for details). Error bars are statistical uncertainties while the thick solid line is the best fit obtained taking into account ^1G and ^3F components only (dash dotted lines). The thin full line is the non-coincidence Auger spectrum measured at the same time and by the same bimodal analyzers used to collect the AR-APECS signal.

are revealed. To verify this hypothesis, we have chosen two different experimental configurations (configurations A and B) of the analyzers. Simulations performed with the aforementioned MSCD code indicate that by selecting one of the two configurations, the relative relevance of the $m_{e_1} = m_{e_2} = 0$ and $m_{e_1} = 0, m_{e_2} = 1$ final substates changes. In particular, in configuration A where contribution from $m_{e_1} = m_{e_2} = 0$ components is dominant, the ^3F term should not contribute to the AR-APECS Auger lineshape. The experiment has confirmed such a prediction, indicating variations of the intensity ratio $^1G/^3F$ as function of the detection angles; in particular, $^1G/^3F$ is larger in configuration A than in B, as expected. This effect is clearly shown in Figure 14.4 where the coincidence M_3VV spectrum measured in configuration A, that enhances contributions from $m_{e_1} = m_{e_2} = 0$ is reported. The measured lineshape is well accounted for by ^1G and ^3F contributions only (peaks shown by dash dotted line in the figure) plus a modest monotonic multiple losses background. The non-coincidence spectrum, as measured together with the coincidence one, is also shown in the figure. The calculated intensities for ^1G and ^3F components are found to be in excellent agreement with the measured non-coincidence spectrum, while the coincidence spectrum yields a ratio larger than 4. A similar discrepancy between coincidence and non-coincidence $^1G/^3F$ ratios has been found for the four different configurations investigated. Moreover, by adopting the simple two-step model described in the previous section, we predict ratios of intensity that, within error bars, always match the experiment [15]. We can then conclude that AR-APECS actually introduces a selectivity in the intermediate core hole state sublevels that allows for discrimination of individual components of the multiplet split Auger spectrum that are otherwise overlapping.

Correlation between the two electrons has also been observed by the energy shift of the Auger electrons in the coincidence acquisition. In the experiment reported in Figure 14.4, the photoelectron analyzer was tuned to an energy 2 eV higher than the $3p_{3/2}$ transition. This avoids any contribution in the AR-APECS spectrum from the $3p_{1/2}$ core state and, at the same time, shows that in this experiment the energy is conserved collectively by the electrons pair rather than by the photo and Auger electron separately. The peak intensity for the ^1G transition is indeed observed at an energy lower than the corresponding non-coincidence transition. The Auger energy shift matches the photoelectron effective distribution detuning (which is, however, less than 2 eV), as expected and already observed [13].

14.3.3 Surface Sensitivity

Enhanced surface sensitivity of electron–electron coincidence experiments has been claimed since early works in this field and a model has been developed to account for it [16]. But it was not until recently that experimental evidences have been acquired [11, 17].

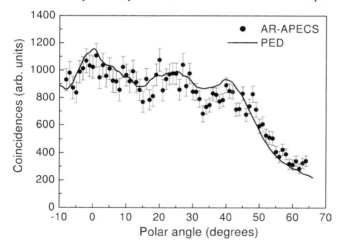

Figure 14.5: Data points are AR-APECS measured in integrated angular mode on the Cu(111) surface; i.e. the angular distribution of the $3p_{3/2}$ photoelectron (measured by bimodal analyzers) in coincidence with M_3VV Auger electron (see text for details). The axial plane is rotated by $35°$ with respect to the X axis (see Figure 14.2). The photon beam impinges onto the surface at a $3°$ grazing angle. Zero of the bimodal angle scale corresponds to the surface normal. Errors bars are statistical uncertainties on the coincidence events. The full line is the non-coincidence photoelectron angular distribution (PED) as measured together with the coincidence signal.

This AR-APECS feature has been studied by an AR-APECS experiment where the $2p_{3/2}$ photoelectron (kinetic energy 160 eV) angular distribution is measured in coincidence with the M_3VV Auger electron (kinetic energy 55 eV) on the Cu(111) surface. We have measured the Cu $2p_{3/2}$ photoelectron polar angle distribution, emitted from the Cu(111) sample on the

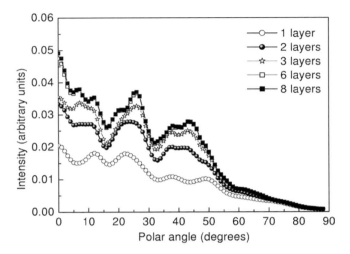

Figure 14.6: MSCD simulations of the $3p_{3/2}$ PED. The polar angle distributions are simulated for several different effective target thicknesses (layers) while the cluster size is kept fixed.

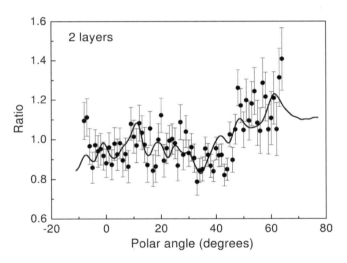

Figure 14.7: Data points are the ratio AR-APECS vs. experimental PED, error bars are statistical uncertainties. Full lines are ratios of MSCD simulation performed for two vs. eight layers effective target thickness.

$(-1, -1)$ direction, in single and in coincidence with the M_3VV Auger electron. The different slopes of these two angular distributions are evident in Figure 14.5 and can be explained by the enhanced surface sensitivity for the coincidence measurements. To account for diffraction effects of multiple photoelectron scattering, we have used the MSCD code with several different effective target thicknesses and the results are reported in Figure 14.6. These sim-

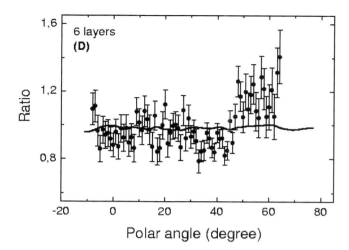

Figure 14.8: Data points are the ratio AR-APECS vs. experimental PED, error bars are statistical uncertainties. Full lines are ratios of MSCD simulation performed for six vs. eight layers effective target thickness.

ulations clearly show that the amplitude of the PED modulation due to diffraction is reduced when the target effective thickness is reduced. Then we shall assume that the experimental ratio between coincidence and non-coincidence photoelectron angular distributions must be sensitive to the difference in effective target thickness experienced by AR-APECS and PED. To put this finding on firmer grounds, in Figures 14.7 and 14.8 we compare this experimental ratio with the ratios computed for different PED emitter depths, using the latter as a trial function to fit the former. Clearly, the best agreement between experimental and calculated ratios is obtained for an emitter depth of two layers, while simulations performed with deeper emitters leads to trial functions that are unacceptable when tested by a maximum likelihood criterion. In this way we have determined, for the first time, the effective sample thickness for an AR-APECS experiment and found it to be as small as two atomic layers.

14.4   Conclusions

In summary, the series of AR-APECS studies performed on a variety of solid targets and under several experimental modes, mainly Auger angle or energy selected, has demonstrated that this novel tool allows one to perform Auger spectroscopy with unprecedented discrimination. The enhanced surface sensitivity, i.e. an effective target thickness close to a few atomic layers, sets AR-APECS among electron spectroscopies with atomic scale spatial resolution, even though limited to the sample depth. Of even wider relevance is the AR-APECS capability to probe the core hole/Auger decay mechanism in the solid with a unique l and m_l selectivity. The Ge and Cu experiments demonstrate that AR-APECS, by properly selecting the kinematics, allows one to study the decay of individual sets of core hole magnetic substrates in individual sets of double hole final state magnetic sublevels. Finally we note that AR-APECS, by detecting the coincident photoelectron, preserves the chirality of the ionization event and then opens the possibility of measuring dichroic effects in the Auger emission, thus providing a possible novel tool for the study of surface magnetic phenomena.

Acknowledgments

The authors are grateful to the ALOISA beamline staff members for the valuable support provided during the AR-APECS experiments performed at the ELETTRA synchrotron radiation facility, and they are also indebted to INFM for financial support provided through the "Supporto ELETTRA" program. Two of us (H.Y. and R.A.B.) acknowledge support by the NSF under Grant No. DMR98-01681.

References

[1] G. Stefani, "Recent Advances in Electron-Electron Coincidence Experiments" in *New Directions in Atomic Physics*, edited by C. T. Whelan, R. M. Dreizler, J. H. Macek, and H. R. J. Walters, Kluwer Academic/Plenum Publishers, New York, 1998.
[2] N. M. Kabachnik, J. Phys. B: At. Mol. Opt. Phys. 25 (1992) L389.
[3] U. Fano and G. Racah, *Irreducible Tensorial Sets*, Academic Press, New York, 1959.
[4] V. V. Balashov, A. N. Grum-Grzhimailo, and N. M. Kabachnik *Polarization and Correlation Phenomena in Atomic collisions*, Kluwer Academic/Plenum Press, New York, 2000.
[5] Y. U. Idzerda, Surf. Rev. Lett. 4 (1997) 161.
[6] D. E. Ramaker, H. Yang, and Y. U. Ydzerda, J. Electron Spectrosc. Relat. Phenom. 68 (1994) 64.
[7] Y. Chen and M. A. Van Hove, *Mscd Package Overview*, Lawrence Berkeley National Laboratory, 1997.
[8] J. J. Rehr and R. C. Albers, Phys. Rev. B 41 (1990) 8139.
[9] F. Hermann and S. Skillman, *Atomic Structure Calculations*, Prentice Hall Inc., Englewood Cliffs, New Jersey, 1963.
[10] R. Gotter, A. Ruocco, A. Morgante, D. Cvetko, L. Floreano, F. Tommasini, and G. Stefani, Nucl. Instr. and Meth. A, 467 (2001) 1468.
[11] R. Gotter, A. Ruocco, M. T. Butterfield, S. Iacobucci, G. Stefani, and R. A. Bartynski, Phys. Rev. B 67 (2003) 033303.
[12] E. Antonides, Phys. Rev. B 15 (1977) 1669.
[13] E. Jensen, R. A. Bartynski, S. L. Hulbert, E. D. Johnson, and R. Garrett, Phys. Rev. Lett. 62 (1989) 71.
[14] R. Kleiman, private communication.
[15] G. Stefani, R. Gotter, A. Ruocco, F. Offi, F. Da Pieve, S. Iacobucci, A. Morgante, A. Verdini, A. Liscio, H. Yao, and R. A. Bartynski, submitted to J. Electron Spectrosc. Relat. Phenom. (2004).
[16] W. S. M. Werner, H. Störi, and H. Winter, Surf. Sci. 518 (2002) L569.
[17] A. Liscio, R. Gotter, A. Ruocco, S. Iacobucci, A. G. Danese, R. A. Bartynski, and G. Stefani, J. Electron Spectrosc. Relat. Phenom. in press (2004).

15 Auger–Photoelectron Coincidence Spectroscopy Studies from Surfaces

S.M. Thurgate, Z.-T. Jiang, G. van Riessen, and C. Creagh

15.1 Introduction

Auger photoelectron coincidence spectroscopy (APECS) from surfaces was first demonstrated in 1979 [1, 2]. Despite this early illustration of the utility of the technique, there have been only a small number of experiments following from this. Perhaps the most significant reason for this has been the perceived difficulty in acquiring data in a reasonable time [3]. Nevertheless, progress has been made, with new instruments appearing on synchrotrons and a new generation of detectors have opened up opportunities for dramatically improved performance.

This chapter will review recent developments in APECS in our laboratory and the progress in understanding that has arisen from these experiments. It will highlight where APECS offers unique opportunities to study processes and point to exciting new directions that are becoming apparent as both the instrumentation and our understanding improve.

15.2 APECS Experiments

The fundamental idea of APECS is that an Auger electron emitted from a photo-ionization process will only be recorded if the photoelectron emitted from the same process is also detected. This can result in significant reductions in the apparent complexity of Auger lines as often the Auger line has a range of contributing ionization processes.

Processes that can produce a core hole that can be filled by an Auger emission, but where the resultant line-shape might be changed, include Auger cascade processes where the original ionization was of a deeper level, Auger final-state shake processes where more than one electron is emitted with the outgoing Auger electron, initial state shake processes where an electron is emitted with the outgoing photoelectron in the creation of the initial state, and secondary ionization from energetic electrons as they escape the solid. This last process is less likely for deep core levels.

The complexity that this system of possible events creates is enhanced by the dynamics of how the resultant holes interact. If they hop away quickly, then the resultant Auger line shape may well reflect, in part, the density of states of the material. If they stay localized, then the line shape will reflect the character of the atom. This distinction between band-like and atom-like Auger lines is well known. APECS provides an opportunity to study these many body phenomena with clarity.

Correlation Spectroscopy of Surfaces, Thin Films, and Nanostructures. Edited by Jamal Berakdar, Jürgen Kirschner
Copyright © 2004 Wiley-VCH Verlag GmbH & Co. KGaA, Weinheim
ISBN: 3-527-40477-5

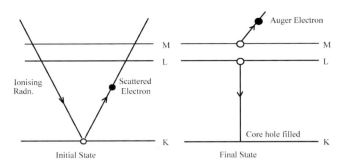

Figure 15.1: A simple Auger process. In an APECS experiment only the electrons that come from this ionization will be counted. Contributions from other ionization processes that lead to a hole in this core level are ignored.

Haak and Sawatzky [4] were the first to demonstrate the power of this approach. The L_3VV Auger spectrum of copper is known to contain a peak, some 2 eV below that of the main ^1G term, which is not accounted for as an atomic term. The feature is shown in Figure 15.2. Roberts et al. [5] correctly identified the source of this as due to a spectator hole that remained local following the Coster–Kronig decay L_2L_3V. The additional electrostatic term slowed the outgoing Auger electron resulting in the 2 eV shift. Haak and Sawatzky were able to show that this peak appeared in coincidence with the $2p_{1/2}$ photoelectron and thus to confirm the origin of the feature.

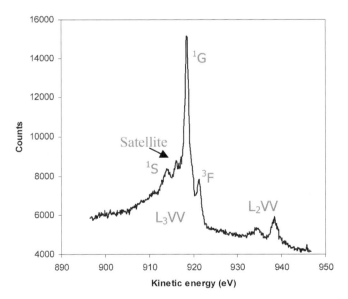

Figure 15.2: High resolution spectra of the LVV Auger line of Cu, showing the ^1G satellite as a consequence of the Auger cascade L_2VV-L_3VV.

The Sawatzky experiment was based on commercial electron analyzers, though modified to increase collection angles. Several years later a group at the National Synchrotron Light Source (NSLS) at Brookhaven National Laboratories, NY developed an APECS system that took advantage of the pulsed nature of the synchrotron light to achieve excellent timing resolution [6]. They have continued to examine a range of surfaces with this facility since then [7]. Further APECS experiments have come on line at synchrotrons in the past few years. These include the multi-analyzer facility at Elettra in Trieste and a new dual coaxial CMA system at the Photon Factory in Japan.

At Murdoch University we have designed and built several APECS systems. Our first analyzer used a slanted channel plates setup to reduce the degradation in timing resolution caused by differences in flight times through various possible flight paths through the analyzer [8]. We have subsequently improved on this apparatus through modifications to the electron analyzers and the electron lenses. Our current system uses two $180°$ hemispherical electron analyzers with five element lenses [9]. Each analyzer has a large separation between the hemispheres and the lens with a wide acceptance angle. A wider acceptance angle increases the probability of detection and so the rate of detection of coincidences [10].

15.3 Applications

15.3.1 Broadening of Cu $2p_{3/2}$

As well as the conventional APECS experiment, it is also possible to perform a "reverse" experiment, where the Auger analyzer is fixed on an Auger peak while the photoelectron analyzer sweeps the correlated photoelectron line. In this way it is possible to isolate the sources of a particular feature in the Auger spectrum. We used this technique to look at the $Cu2p_{3/2}$ photoelectron line of Cu in coincidence with terms in the L_3VV Auger line.

The L_3VV Auger line of Cu is atom-like, with peaks resembling the atomic terms seen in spectra from Cu vapor. This phenomenon was first explained independently by Cini and Sawatzky in 1977 [11, 12]. This means that it is possible to resolve the Cu L_3VV spectra into spectroscopic terms. The main terms are labeled in Figure 15.2. We collected coincidence photoelectron spectra of the $2p_{3/2}$ line with the ^1G and ^3F terms (Figures 15.3 and 15.4).

In coincidence with the ^1G term, the line narrows on the low kinetic energy side, but the centroid is unshifted. With the ^3F term, the line shifts 0.15 eV towards lower kinetic energies. This was a very puzzling result at first. The simplest views of the Auger process suggest that the Auger emission is independent of the manner in which the core hole is produced. That is clearly not so in this case. However, a full account of this phenomenon was contingent on a more detailed description of the APECS line-shape. In the past few years Ohno has developed a many-body approach to explaining the coincidence line-shape [13, 14]. The approach he has taken is to observe that the coincidence line-shape is a consequence of the decay of a single core hole, and the dynamics of that process influence the shape of both the coincidence photoelectron line and the coincidence Auger line. He has been able to show that the singles spectrum is essentially proportional to the spectral function of the initial core hole, while the coincidence line shapes depend also on the imaginary part of the initial core hole self energy and density of final states.

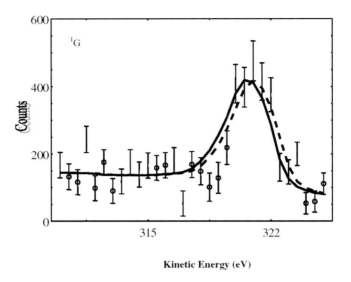

Kinetic Energy (eV)

Figure 15.3: Cu $2p_{3/2}$ in coincidence with the L_3VV ^1G term. The solid line shows the singles data while the dashed line shows the best fit to the shifted coincidence data.

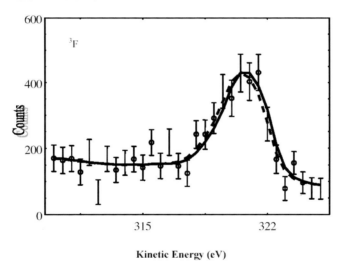

Kinetic Energy (eV)

Figure 15.4: Cu $2p_{3/2}$ in coincidence with the L_3VV ^3F term. The solid line shows the singles data while the dashed line shows the best fit to the shifted coincidence data.

The case he has concentrated on is one where the initial core hole can decay either directly or indirectly through the excitation of a shake up/off electron. He points out that in such a case an APECS measurement of the photoelectron line will contain contributions from both the direct transition and the indirect one via the shakeup/down process and under these conditions the coincidence photoelectron line can move and change width, as observed in the case of Cu L_3VV.

15.3.2   Broadening of Ag 3d$_{5/2}$

The test of this proposal was to see if the Ag 3d$_{5/2}$ photoelectron line moved relative to the singles line in coincidence with different terms of the M$_{4,5}$VV Auger line. The shakeup/down satellite of the Ag 3d$_{5/2}$ is known to be small compared to that of Cu 2p$_{3/2}$. Ohno's theory predicted therefore that the photoelectron line would not shift in coincidence with different terms of the Ag M$_{4,5}$VV line. A high resolution XPS spectrum of Ag M$_{4,5}$VV, with terms marked, is shown in Figure 15.5.

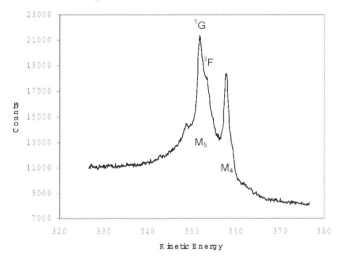

Figure 15.5: High resolution XPS spectrum of the Ag M$_{4,5}$VV Auger line.

The coincidence photoelectron spectra from the 3d$_{5/2}$ in coincidence with the ^{1}G and ^{3}F terms are shown in Figures 15.6 and 15.7 respectively. It can be seen that the shift in these peaks is very much less than in the corresponding case for Cu 2p$_{3/2}$. The shifts for each line are shown in Table 15.1.

Material	Auger Line	Shift
Cu	^{1}G L$_3$VV	+0.25
	^{3}F L$_3$VV	−0.15
Ag	^{1}G M$_4$VV	+0.04
	^{3}F M$_4$VV	−0.03

Table 15.1: Measured shifts in the coincidence photoelectron peaks of Cu and Ag with different Auger lines.

15.3.3   Disorder Broadening

It has been known for some time that the photoelectron lines of compositionally disordered alloys can demonstrate a broadening due to the range of chemically different sites that atoms of each component can find themselves in. It was proposed [15] that we measure the M$_{4,5}$VV Auger line of Ag in Ag$_{0.5}$Pd$_{0.5}$ alloy in coincidence with the 3d$_{5/2}$ line. Ag$_{0.5}$Pd$_{0.5}$ is known

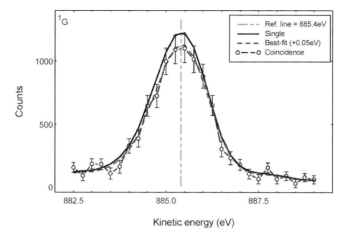

Figure 15.6: Ag $3d_{5/2}$ Photoelectron line in coincidence with the ^{1}G term of the M_5VV Auger line. The solid line is the singles data, while the best fit of the singles lineshape to the coincidence is shown as a dashed line.

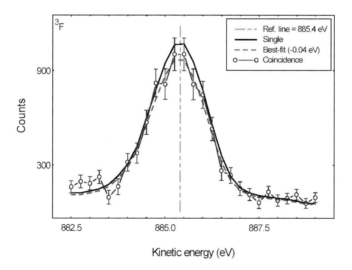

Figure 15.7: Ag $3d_{5/2}$ Photoelectron line in coincidence with the ^{3}F term of the M_5VV Auger line. The solid line is the singles data, while the best fit of the singles lineshape to the coincidence is shown as a dashed line.

to be compositionally disordered with broadened photoemission lines. We then shifted the central position of the photoelectron analyzer within the broadened width to see if there was a corresponding movement of the Auger line. The data from the experiment is shown in Figures 15.8, 15.9 and 15.10, and is summarized in Table 15.2.

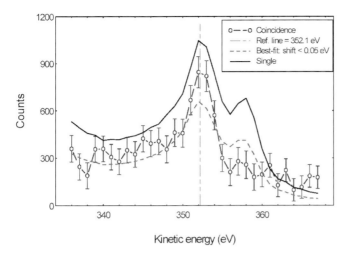

Figure 15.8: Ag M_5VV with $3d_{5/2}$ at the center of the photoemission line. There is no shift of the coincidence Auger line relative to the singles.

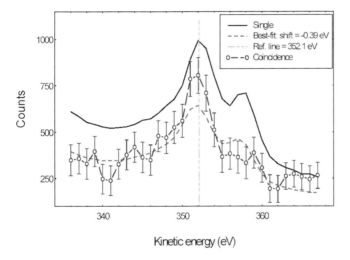

Figure 15.9: Ag M_5VV with $3d_{5/2}$ 0.8 eV above the center of the photoemission line. The coincidence Auger line has shifted -0.39 eV.

The incident radiation was Mg-Kα (not monchromatized) with a line-width of 700 meV. Thus the movement of the Auger line is due to the selection by the photoelectron analyzer of ionization events where the binding energy is either greater than or less than the mean energy. The data show convincingly that when the binding energy is less than the mean, the Auger energy is greater than the mean, and vice versa. This confirms the picture that this alloy is compositionally disordered with atoms placed in a range of locations of different binding energy.

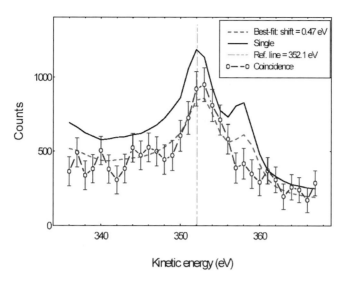

Kinetic energy (eV)

Figure 15.10: Ag M_5VV with $3d_{5/2}$ 0.8 eV below the center of the photoemission line. The coincidence Auger line has shifted 0.47 eV.

Photoemission Analyzer Setting Relative to Center of Singles / eV	Measured Movement of M_5VV Auger Line Relative to Singles / eV
0.0	0.0
+0.8	−0.39
−0.8	+0.47

Table 15.2: Changes in the photoelectron analyzer energy settings and the corresponding measured movement of the Auger line.

15.4 Conclusions

APECS is finding application in understanding a wide range of problems. In this chapter we have shown how it has been used to investigate differences in decay mechanisms in Cu and Ag and how it can be used to probe the nature of disordered broadening. There has been progress in understanding that the coincidence line-shape is not the simple sum of the component parts, but rather it can reflect the dynamics of the core hole decay processes.

References

[1] H.W. Haak, G.A. Sawatzky, and T.D. Thomas, Phys. Rev. Lett. **41** (1978), 1825.

[2] H.W. Haak, University of Gronigen, PhD Thesis. (1983).

[3] S.M. Thurgate, J. Electron Spectrosc. Relat. Phenom. **100** (1999) 161.

[4] G.A. Sawatzky, Treatise Mater. Sci. Technol. **30** (1988) 167.

[5] E.D. Roberts, P. Weightman, and C.E. Johnson, J. Phys. C **8** (1975) 301.

[6] E. Jensen, R.A. Bartynski, S.L. Hulbert, E.D. Johnson and R. Garrett, Phys. Rev. Lett. **62** (1988) 71.

[7] D.A Arena, R.A. Bartynsk, R.A. Nayak, A.H. Weiss, S.L. Hulbert, and M. Weinert, Phys. Rev. Lett. **91** (2003) 176, 403.

[8] S.M. Thurgate, B. Todd, B. Lohmann, and A Stelbovics, Rev. Sci. Instrum. **12** (1990) 3733.

[9] S.M. Thurgate, C.P. Lund, C. Creagh, and R. Craig, J. Electron Spectrosc. Relat. Phenom. **93** (1998) 209.

[10] S.M. Thurgate, Surf. Interface Anal. **20** (1993) 627.

[11] M. Cini, Solid State Commun. **24** (1977) 681.

[12] G.A. Sawatzky, Phys Rev. Lett. **39** (1977) 504.

[13] M. Ohno, Phys. Rev. B **58** (1998) 12795.

[14] M. Ohno , J. Electron Spectrosc. Relat. Phenom. **124** (2002) 53.

[15] Z.-T. Jiang, S.M. Thurgate, G. van Riessen P. Wilkie, and C. Creagh, J. Electron Spectrosc. Relat. Phenom., **130** (2003) 33.

16 Development of New Apparatus for Electron–Polar-Angle-Resolved-Ion Coincidence Spectroscopy and Auger–Photoelectron Coincidence Spectroscopy

Kazuhiko Mase, Eiichi Kobayashi, and Kouji Isari

Coincidence measurement of energy-selected electrons and mass-selected ions (electron–ion coincidence (EICO) spectroscopy) is a powerful technique to clarify the ion desorption mechanism induced by electron transitions, because excited states leading to ion desorption are directly identified. Information on the coincidence ions, however, has been limited so far to the mass and the yield. In order to obtain information on the kinetic energy and the desorption polar angle of ions, we have developed a new electron–polar-angle-resolved-ion coincidence apparatus, which consists of a coaxially symmetric mirror electron energy analyzer and a miniature polar-angle-resolved time-of-flight ion mass spectrometer ((TOF-MS)). The TOF-MS consists of a shield for the electric field, an ion drift electrode with three meshes, and microchannel plates with three concentric anodes. By using SIMION 3D version 7.0 we simulated ion trajectories of H^+ for the TOF-MS with a drift bias of -30 V. The results show that the desorption angles of H^+ with a kinetic energy of 3 eV detected by the innermost anode, the intermediate anode, and the outermost anode are $0°-17°$, $22°-48°$, and $57°-90°$, respectively. By assembling a miniature cylindrical mirror electron energy analyzer (CMA) with a diameter of 26 mm in a coaxially symmetric mirror analyzer coaxially and confocally we have developed an apparatus for Auger–photoelectron coincidence spectroscopy (APECS). The CMA consists of a shield for the electric field, inner and outer cylinders, a pinhole, and an electron multiplier. The performance was tested by measuring Si LVV Auger–Si 2p photoelectron coincidence spectra of a Si(111) surface. Features of the APECS apparatus are as follows: 1) Coincidence signal detection efficiency is improved by one order of magnitude from previous ones because of the large solid angle of the coaxially symmetric mirror analyzer and the CMA. 2) Positioning is quite easy, because the coaxially symmetric mirror analyzer and the CMA are assembled confocally on a rod with a mechanism for xyz positioning and tilt adjustment. 3) It can be installed in a multi-purpose ultrahigh vacuum chamber because it is constructed on a 203 mm-outer-diameter conflat flange with a 50 mm retractable mechanism. 4) The production cost is low because the structure is simple and the number of parts is relatively small.

16.1 Introduction

Desorption induced by electronic transitions (DIET) is an active research field in surface science [1–3]. Especially, the study of Auger stimulated ion desorption (ASID, Figure 16.1)

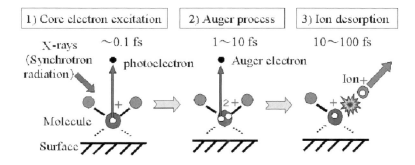

Figure 16.1: Diagram of Auger stimulated ion desorption [46]. When a surface is irradiated by X-rays or synchrotron radiation ion desorption is induced by the following three-step processes, i.e., 1) formation of a core-hole by a core-electron emission ($\sim$0.1 fs), 2) formation of a two-hole state by an Auger transition (1−10 fs), 3) ion desorption induced by Coulomb repulsion of the two holes and by electron missing from the bonding orbitals (10−100 fs).

induced by core-electron transitions has developed rapidly owing to the development of synchrotron radiation facilities [4–7]. For ion desorption studies energy-selected-electron–mass-selected-ion coincidence (EICO) spectroscopy is a powerful tool because it offers ion mass spectra for selected core excitations or selected Auger transitions [8, 9].   In 1985 Knotek and Rabalais developed an EICO apparatus combined with an electron beam [10]. The coincidence signal, however, was not clear enough to discuss the ion desorption mechanism. EICO spectroscopy was not applied to surface studies until 1996, because of several problems characteristic to surfaces, such as the abundance of secondary electrons and the high probability of recapture or neutralization of ions. In 1996, Mase and collaborators developed an improved EICO apparatus implemented with an electron beam [11] and synchrotron radiation [12]. Synchrotron radiation is more advantageous than an electron beam because the resonant excitations are accessible and the secondary electrons are drastically reduced. Since then second [13], third [8], and fourth [14] EICO analyzers have been developed. The performance, such as coincidence signal to background ratio and energy resolution, has been improved at every remodeling.

EICO spectroscopy has now been widely applied to studies of ion desorption from condensed molecules such as H_2O [8, 15–21], NH_3 [22–24], CH_3CN [25–27], $Si(CH_3)_4$ [28], C_6H_6 [29] and for molecules showing site-specific fragmentation [8, 30–35]. The EICO studies of condensed molecules have been described in detail in a previous review [9].   EICO has also been applied to a poly-methylmethacrylate (PMMA) thin film [26, 36–39], a Si(100) surface terminated by fluorine [13], a Si(111) fluorinated by XeF_2 [40], a H_2O dissociatively chemisorbed on a Si(100) surface (HO/Si(100)) [19, 21, 41, 42], a $CaF_2(111)$ film epitaxially grown on Si(111) [13, 43] and a $TiO_2(110)$ surface [44, 45]. Further information on EICO studies of surfaces is available at the homepage of http://pfwww.kek.jp/eico/EICO.html. In this review we describe details of the latest (fourth) EICO analyzer [14, 46] in Section 16.2 and a new electron–polar-angle-resolved-ion coincidence analyzer [47] in Section 16.3.

Using the technique developed for the EICO apparatus, we have constructed a new apparatus for Auger–photoelectron coincidence spectroscopy (APECS). Advantages of APECS for studies of Auger processes are described in reviews [48–51] and other chapters of this book. APECS is also necessary to study DIET because minor Auger processes often play an important role in DIET. In a pioneering work of O^+ desorption induced by Ti 3p core-level ionization at a TiO_2 surface, Knotek and Feibelman concluded that O^+ desorption is induced by interatomic shake-off Auger processes [52]. EICO studies, however, suggest that charge transfer (CT) induced by core-level photoemission and Auger is responsible for O^+ desorption from TiO_2 [44, 45]. Based on the theory of Kotani and Toyozawa [53] Uozumi showed that CT induced by core-level photoemission and Auger forms three-hole states that are expected to result in O^+ desorption from TiO_2 [54]. Shake-up core-level ionization and shake-up Auger were attributed as the cause of H^+ desorption from HO/Si(100) [19,21,41,42]. APECS measurement for the specific site taking advantage of core-chemical shift will clarify the mechanism of excitation site-specific ion desorption [8, 30–35, 55]. We describe details of a new APECS analyzer [56] in Section 16.4.

16.2 EICO Analyzer Using a Coaxially Symmetric Electron Energy Analyzer and a Miniature Time-of-Flight Ion Mass Spectrometer (TOF-MS)

16.2.1 Coaxially Symmetric Electron Energy Analyzer [46]

The latest (fourth) EICO apparatus consists of a new coaxially symmetric mirror electron energy analyzer and a miniature time-of-flight ion mass spectrometer (TOF-MS). The original coaxially symmetric mirror analyzer was developed by K. Siegbahn et al. in 1997 [57]. The potential (φ) of the coaxially symmetric electric field is described by the Laplace equation

$$\frac{1}{r}\frac{\partial}{\partial r}r\frac{\partial \varphi}{\partial r} + \frac{\partial^2 \varphi}{\partial z^2} = 0 \tag{16.1}$$

where r is the distance from the axis and z is the coordinate along the direction of the axis [57]. Then the coaxially symmetric potential is given by Eq. (16.2).

$$\varphi = a\ln r + b\left(\frac{r^2}{2} - z^2\right) + cz + d \tag{16.2}$$

The cylindrical mirror electron energy analyzer (CMA) corresponds to the case of $b = c = 0$,

$$\varphi = a\ln r + d \tag{16.3}$$

and the inner and outer electrodes take the form of a cylinder. Siegbahn et al. found that the convergence of the electron trajectory is better in the electric field given by $b = -a$ and $c = 0$

$$\varphi = a\ln r - a\left(\frac{r^2}{2} - z^2\right) + d \tag{16.4}$$

than that of a CMA, and developed a coaxially symmetric mirror analyzer that consists of the inner and outer electrodes [57]. The performance and characteristics of Siegbahn's analyzer are as follows: 1) Solid angle: $0.94-0.19$ sr. 2) Energy resolution: $E/\Delta E = 200-4000$ (FWHM). 3) Electron energy range: $0-3000$ eV. 4) An electron gun is installed. 5) Constructed on a conflat flange with an outer diameter of 253 mm. The analyzer, however, had a weak point in that the performance was degraded by the disturbance of the electric field near the end plates. We have improved their analyzer and developed a modified one, which consists of an inner electrode, an outer electrode, three sets of compensation electrodes, a pinhole with a diameter of 0.8 mm, microchannel plates (MCP, Hamamatsu Photonics, F4655), and a magnetic shield (Figures 16.2 and 16.3) [46]. The trajectories of electrons and isoelectric lines simulated with SIMION 3D version 7.0 (http://www.sisweb.com/simion.htm) are also shown in Figure 16.2. Figure 16.4 shows electron trajectories around the pinhole of the coaxially symmetric mirror analyzer. The resolution of the analyzer was estimated at UVSOR BL-2B1 by measuring an Au 4f photoelectron spectrum of a gold film (Figure 16.5). The performance and characteristics of our analyzer are as follow. 1) Solid angle: 1.2 sr. 2) Energy resolution: $E/\Delta E \sim 130$ (FWHM). 3) Electron energy range: $0-3000$ eV. 4) A miniature (TOF-MS) can be installed. 5) Constructed on a conflat flange with an outer diameter of 203 mm. The distance between the sample and the flange is 230 mm.

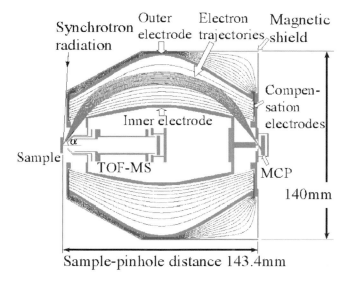

Figure 16.2: Cross section of the coaxially symmetric mirror electron energy analyzer [46]. The analyzer consists of an inner electrode, an outer electrode, compensation electrodes and a magnetic shield. Microchannel plates (MCP) are used to detect electrons. A time-of-flight mass spectrometer (TOF-MS) can be installed coaxially inside the analyzer for application to EICO spectroscopy. The isoelectric lines simulated by the SIMION 3D version 7.0 are shown for the inner electrode and the magnetic shield setting at 0 V, and the outer electrode at -100 V. Electron trajectories are also shown for the electron kinetic energy of 182.75 eV and the emission polar angles (α) of $51.7°-67.5°$ (the solid angle is 1.2 sr).

16.2.2   Miniature Time-of-Flight Ion Mass Spectrometer (TOF-MS) [47]

Figure 16.6 shows the miniature time-of-flight ion mass spectrometer (TOF-MS) we developed for the EICO apparatus [47]. It consists of a shield for the electric field, an ion-extraction electrode, a drift tube, and microchannel plates (MCP). The transmittance of the three meshes inserted perpendicular to the axis of the TOF tube is 0.47, and the ion detection efficiency of the MCP is 0.60. The distance between the sample and the front end of the (TOF-MS) is 3.4 mm. The TOF-MS is so compact that it would be useful also for other fields such as mass spectroscopy in the future landed planetary missions [58]. Figure 16.7 shows TOF contour maps of H^+ for the miniature TOF-MS with a single anode as a function of the desorption polar angle and the kinetic energy [47]. The miniature TOF-MS was installed coaxially in the coaxially symmetric mirror analyzer as shown in Figure 16.2.

16.2.3   EICO Apparatus Using a Coaxially Symmetric Mirror Analyzer and a Miniature TOF-MS [14]

We developed the fourth EICO apparatus, which consists of a coaxially symmetric mirror analyzer, a miniature (TOF-MS), and a positioning mechanism [14]. The incidence angle of the synchrotron radiation is $84°$ from the surface normal. The glancing incidence is advantageous for studies of desorption induced by p-polarized radiation [59]. The fourth EICO apparatus was used for investigations of fluorination of Si(111) by XeF_2 [40].

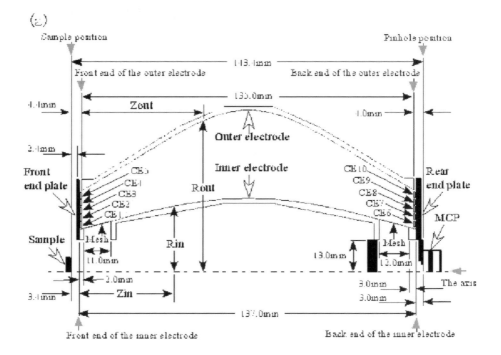

(b)

Outside coordinates of the inner electrode (mm)				Inside coordinates of the outer electrode (mm)							
Zin	*Rin*	*Zin*	*Rin*	*Zout*	*Rout*	*Zout*	*Rout*	*Zout*	*Rout*	*Zout*	*Rout*
0.0	17.2	70.5	30.1	0.0	33.2	31.1	52.1	85.1	62.8	119.1	43.3
21.6	23.0	79.5	29.7	0.7	33.6	32.7	53.1	87.3	61.8	120.6	42.4
22.8	23.3	83.9	29.3	2.4	34.6	34.3	54.0	89.4	60.9	122.2	41.4
24.4	23.7	87.3	28.9	4.1	35.5	35.9	55.0	91.3	59.9	123.8	40.4
26.0	24.1	90.2	28.5	5.7	36.5	37.6	56.0	93.2	58.9	125.4	39.4
27.6	24.5	92.8	28.1	7.3	37.5	39.3	57.0	95.0	57.9	127.0	38.5
29.3	24.9	95.2	27.7	9.0	38.5	41.0	57.9	96.7	57.0	128.7	37.5
31.0	25.3	97.4	27.3	10.6	39.4	42.8	58.9	98.4	56.0	130.3	36.5
32.8	25.7	99.5	26.9	12.2	40.4	44.7	58.9	100.1	55.0	131.9	35.5
34.7	26.1	101.4	26.5	13.8	41.4	46.6	60.9	101.7	54.0	133.6	34.6
36.6	26.5	103.3	26.1	15.4	42.4	48.7	61.8	103.3	53.1	135.0	33.8
38.5	26.9	105.2	25.7	16.9	43.3	50.9	62.8	104.9	52.1		
40.6	27.3	107.0	25.3	18.5	44.3	53.3	63.8	106.5	51.1		
42.8	27.7	108.7	24.9	20.1	45.3	56.1	64.7	108.1	50.1		
45.2	28.1	110.4	24.5	21.6	46.3	59.6	65.7	109.7	49.2		
47.8	28.5	112.0	24.1	23.2	47.2	66.6	66.7	111.2	48.2		
50.7	28.9	113.6	23.7	24.8	48.2	69.4	66.7	112.8	47.2		
54.1	29.3	115.2	23.3	26.3	49.2	76.4	65.7	114.4	46.3		
58.5	29.7	116.4	23.0	27.9	50.1	79.9	64.7	115.9	45.3		
67.5	30.1	137.0	17.5	29.5	51.1	82.7	63.8	117.5	44.3		

(c)

	CE1	CE2	CE3	CE4	CE5	CE6	CE7	CE8	CE9	CE10
Inner diameter (mm)	36.0	42.0	48.0	54.0	60.0	37.0	43.0	49.0	55.0	61.0
Outer diameter (mm)	41.0	47.0	53.0	59.0	65.0	42.0	48.0	54.0	60.0	66.0

(d)

	CE1	CE2	CE3	CE4	CE5	CE6	CE7	CE8	CE9	CE10
Inner diameter (mm)	39.0	47.0	55.0	–	–	40.0	48.0	56.0	–	–
Outer diameter (mm)	45.0	53.0	64.0	–	–	46.0	54.0	65.0	–	–

Figure 16.3: The coaxially symmetric mirror analyzer for EICO apparatus [46]. (a) Cross section and scale. CE denotes compensation electrode. (b) Table of the outside coordinates of the inner electrode (Z_{in}, R_{in}) and the inside coordinates of the outer electrode (Z_{out}, R_{out}). Z_{in} and Z_{out} represent the distances from the front end of each electrode, while R_{in} and R_{out} represent the distances from the axis, as shown in (a). Tables of the outer and inner diameters of the CEs are also shown (c) for the analyzer with five sets of CEs, and (d) for the analyzer with three sets of CEs. Z_{in}, Z_{out}, R_{in}, and R_{out} satisfy the following equations: $Z_{in} = 95z_{in} + 69$, $Z_{out} = 95z_{out} + 68$, $R_{in} = 95r_{in}$, $R_{out} = 95r_{out}$, $0 = \ln r_{in} - \left(\frac{r_{in}^2}{2} - z_{in}^2 \right) + 1.2$, $0.6 = \ln r_{out} - \left(\frac{r_{out}^2}{2} - z_{out}^2 \right) + 1.2$.

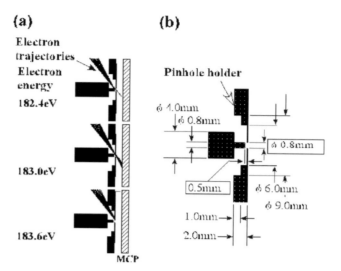

Figure 16.4: (a) Electron trajectories at the pinhole of the coaxially symmetric mirror analyzer [46]. The electron kinetic energies are 182.4, 183.0 and 183.6 eV. (b) Scale around the pinhole of the coaxially symmetric mirror analyzer [46]. The diameter of the pinhole is changed when larger transmittance is required (see Section 16.4.3).

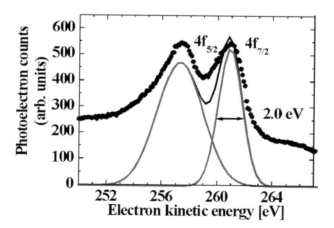

Figure 16.5: Photoelectron spectrum of a gold film in the Au 4f region [46]. Based on Gaussian curve fitting the energy resolution was estimated to be $E/\Delta E = 130$ at the Au $4f_{7/2}$ peak, the full width at half maximum (FWHM) of which was 2.0 eV.

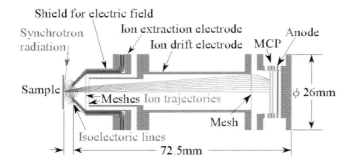

Figure 16.6: Cross section of a miniature TOF-MS with a single anode [47]. The TOF-MS consists of a shield for electric field, an ion extraction electrode with a mesh, an ion drift electrode with two meshes, and microchannel plates (MCP). The isoelectric lines with 3 V step between the sample and the ion extraction electrode, and trajectories of ions from a pointed source with a kinetic energy of 2 eV for desorption polar angles of $0° - 45°$ with $5°$ step are also shown based on the simulation with the SIMION 3D ver. 7.0. The voltage of the sample is -0 V, that of the ion extraction electrode is -30 V, that of the ion drift electrode is -300 V and that of the MCP entrance is -2000 V.

16.3   EICO Analyzer Using a Coaxially Symmetric Mirror Analyzer and a Miniature Polar-Angle-Resolved TOF-MS [47]

16.3.1   Miniature Polar-Angle-Resolved TOF-MS with Three Concentric Anodes [47]

Figure 16.7 shows that information on desorption polar angles and kinetic energy of coincidence ions can be obtained with a position-sensitive detector. Electron stimulated desorption ion angular distribution (ESDIAD) using MCP with an image-readout is widely applied for studies of configurations of surface molecules [3]. The size of such MCP, however, is much larger than the inner diameter of our coaxially symmetric mirror analyzer. So, we developed a miniature (TOF-MS) with three concentric anodes. Figure 16.8 shows the miniature TOF-MS for the EICO apparatus [47]. It consists of a shield for electric field, a drift tube with three meshes, and MCP with three concentric anodes. The outer diameter of the innermost anode (anode 1) is 3.0 mm, the inner and outer diameters of the intermediate anode (anode 2) are 4.0 mm and 8.0 mm, and those of the outer anode (anode 3) are 9.0 mm and 14.5 mm. Figure 16.9 shows TOF contour maps of H^+ for the miniature TOF-MS with a three anode as a function of desorption polar angle and kinetic energy.

In order to avoid disturbing the electric field between the sample and the drift tube the sample holder must have a flat structure. Figure 16.10 shows a holder, which we have developed, for a silicon single-crystalline substrate. The size of the silicon wafer is about $5 \times 25 \times 0.5$ mm^3, and the bump from the silicon surface to the upper and lower tantalum

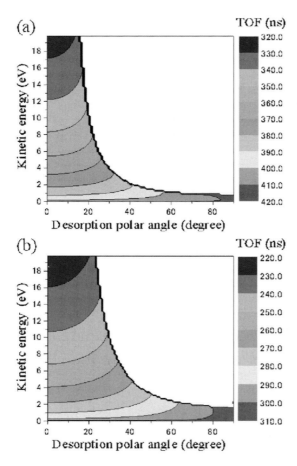

Figure 16.7: TOF contour map of H^+ for the miniature TOF-MS with a single anode as a function of the desorption polar angle and the kinetic energy [47]. The geometry is the same as that of Figure 16.6. The voltage of the sample is 0 V, that of the ion extraction electrode is -30 V, those of the drift electrode are (a) -300 V and (b) -1000 V, and that of the MCP entrance is -2000 V.

electrodes is 0.5 mm in height. To avoid disturbing the electric field around the sample two tantalum plates are set either side of the silicon substrate. The surface can be cleaned by direct heating and can be cooled by liquid nitrogen to 100 K.

16.3.2   Electron–Polar-Angle-Resolved-Ion Coincidence Apparatus [47]

Figure 16.11 shows an electron–polar-angle-resolved-ion coincidence apparatus that consists of a coaxially symmetric mirror analyzer and the miniature polar-angle-resolved TOF-MS with three anodes. The features of the apparatus are as follows: 1) The desorption polar angle and kinetic energy of the ions can be estimated for the selected core-excitation-final-states or the selected Auger-final-state. 2) The configuration of the surface molecules responsible for the coincidence ions can be estimated from the desorption polar angles that reflect the directions of the surface bond. 3) The ion desorption mechanism can be studied for the selected desorption polar angle. Using the apparatus we have studied H^+ desorption induced by $4a_1 \leftarrow O\ 1s$ resonant transitions of condensed H_2O at 100 K [60]. The resolution of the desorp-

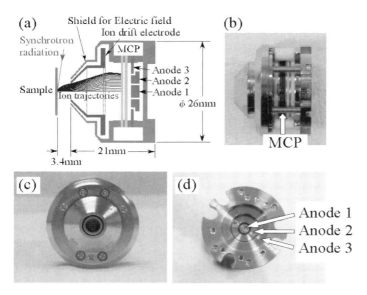

Figure 16.8: (a) Cross section of a polar-angle-resolved miniature TOF-MS with three anodes [47]. The TOF-MS consists of a shield for electric field, an ion drift electrode with three meshes, MCP, and three concentric anodes. Trajectories for ions with a kinetic energy of 2 eV from a pointed source are shown for the desorption polar angles of $0°-90°$ with $5°$ step based on the simulation with SIMION 3D ver. 7.0. The voltage of the sample is 0 V, that of the drift electrode is -30 V and that of the MCP entrance is -2000 V. (b–d) Photographs of the polar-angle-resolved miniature TOF-MS with three concentric anodes [47]. (b) The front view, (c) the side view, and (d) the three concentric anodes.

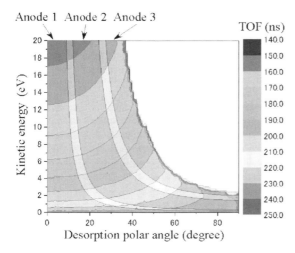

Figure 16.9: TOF contour map of H^+ for the polar-angle-resolved miniature TOF-MS with three anodes as a function of the desorption polar angle and the kinetic energy [47]. The geometry and the electrode voltages are the same as those of Figure 16.8(a).

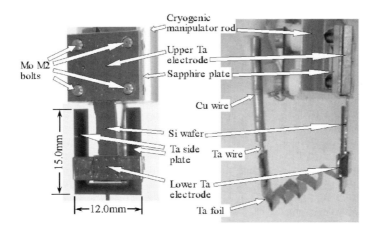

Figure 16.10: Photographs of a silicon wafer holder with Ta electrodes for direct current heating [47].

tion polar angle and of the kinetic energy for the (TOF-MS) with three anodes, however, is not so good. Thus we have developed a miniature TOF-MS with four anodes. Studies of ion desorption using the new apparatus are in progress [61].

16.4   APECS Apparatus Using a Coaxially Symmetric Mirror Analyzer and a Miniature CMA

16.4.1   Introduction

In general, an apparatus for Auger–photoelectron coincidence spectroscopy (APECS) consists of two electron energy analyzers, such as two concentric hemispherical ones [62–64], two double-path CMAs [65], or two $127°$ cylindrical deflecting analyzers [66]. However, there were the following problems. 1) Since the solid angle of the analyzers was small, the detection efficiency of the coincidence signal was low. 2) Positioning of the two analyzers was difficult. 3) A single-purpose ultrahigh vacuum chamber where two analyzers could be installed was necessary. 4) The production cost was high. To overcome these problems we have constructed a new APECS apparatus by assembling a miniature CMA in a coaxially symmetric mirror analyzer coaxially and confocally. Although the CMA is compact ($26 \, \text{mm} \times 71 \, \text{mm}$), it has a wide solid angle of 0.72 sr. We report the details of the APECS apparatus and the result of its application to the Si LVV Auger–Si 2p photoelectron coincidence spectrum of a clean Si(111) surface in the following sections.

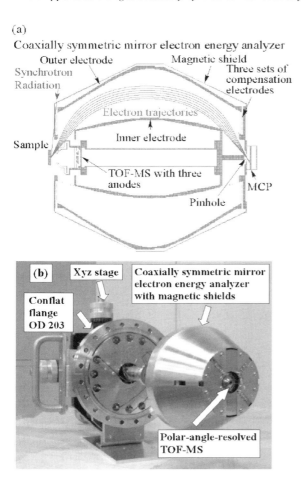

(a)

Coaxially symmetric mirror electron energy analyzer

Figure 16.11: Electron–polar-angle-resolved-ion coincidence analyzer that consists of a coaxially symmetric mirror electron energy analyzer and the miniature polar-angle-resolved TOF-MS with three anodes. (a) Cross section [47] and (b) a photograph [46]. OD stands for outer diameter.

16.4.2 Miniature CMA [56]

Figures 16.12 and 16.13 show the miniature CMA that we developed. It consists of a shield for electric field, inner and outer cylinders, a pinhole with a diameter of 1.0 mm, and an electron multiplier (BURLE, 5901 MAGNUM). We used non-magnetic stainless steel 310S and inconel 600 for the metallic parts, and high purity alumina and kapton foil as insulators. Two meshes with a thickness of 0.05 mm and a transmittance of 0.84 were spotwelded on the inner electrode for the electron path. The trajectories of electrons and isoelectric lines simulated with SIMION 3D version 7.0 are also shown in Figure 16.12. Since the CMA can

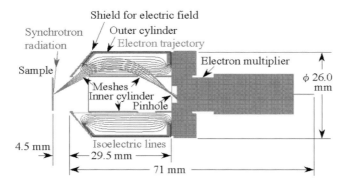

Figure 16.12: Cross section of a miniature cylindrical mirror electron energy analyzer (CMA) [56]. The CMA consists of a shield for electric field, an inner cylindrical electrode with two meshes, an outer cylindrical electrode, and an electron multiplier. The isoelectric lines with 10 V step and trajectories of electrons from a point source with a kinetic energy of 168 eV for polar angles of $29° - 41°$ with $2°$ step are shown based on the simulation with the SIMION 3D version 7.0. The voltages of the sample, the inner cylindrical electrode and the entrance of the electron multiplier are 0 V while that of the outer cylindrical electrode is -100 V.

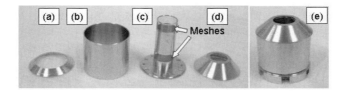

Figure 16.13: Photographs of the parts of the miniature CMA [56]. (a) The cover of the outer cylinder, (b) the outer cylinder, (c) the inner cylinder, and (d) the shield for electric field. (e) The constructed miniature CMA without an electron multiplier.

be constructed on a conflat flange with an outer diameter of 70 mm, it would be useful also for other fields such as photoelectron, Auger electron, and ion spectroscopies [67, 68].

16.4.3   New APECS Apparatus [56]

Figures 16.14 and 16.15 show the new APECS apparatus that we developed [69]. The measurement system is almost the same as that of the EICO. The diameter of the pinhole at the focal point of the coaxially symmetric mirror analyzer was set to 1.2 mm. The analyzers were constructed on a conflat flange with an outer diameter of 203 mm (CF203) combined with a xyz stage (MDC, PSM-1502) and tilt adjustment mechanism. It is 50 mm retractable and its weight is 15.8 kg. Therefore, the focus of the coincidence spectroscope can easily be set to the position of the sample irradiated by synchrotron radiation. Moreover, it can be installed

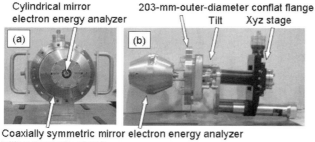

Cylindrical mirror electron energy analyzer

203-mm-outer-diameter conflat flange Tilt Xyz stage

(a) (b)

Coaxially symmetric mirror electron energy analyzer with a magnetic shield

Figure 16.14: Photographs of the Auger–photoelectron coincidence analyzer [56]. (a) The front view and (b) the side view.

into a multi-purpose ultrahigh vacuum chamber with a synchrotron radiation incidence port and a CF203 port.

16.4.4 Application to Auger–Photoelectron Coincidence Spectroscopy [56]

The APECS apparatus was installed in a multi-purpose ultrahigh vacuum chamber at BL12A in the synchrotron radiation facility (Photon Factory, PF) of the Institute of Materials Structural Science. A Si(111) surface cleaned by direct heating in ultrahigh vacuum was used as a sample. The surface was irradiated by p-polarized synchrotron radiation with an incidence angle of $84°$ from the surface normal. The beam spot on the sample was about 2×3 mm^2. Figure 16.16 shows Si 2p photoelectron and Si LVV Auger electron spectra of a clean Si(111) surface measured at $h\nu = 130$ eV. The energy resolution of the coaxially symmetric mirror analyzer and miniature CMA at the electron kinetic energy of 27 eV (Si 2p peak position) was estimated to be 0.7 eV and 2.2 eV, respectively. The cause for the insufficient resolution of the miniature CMA is thought to be the large diameter of the pinhole, the large beam spot on the sample, insufficient magnetic shielding, the large spacing of the mesh, etc. Figure 16.17 shows the time-of-flight difference spectrum of Si LVV Auger electrons with a kinetic energy of 91 eV triggered by Si 2p photoelectrons with a kinetic energy of 26.9 eV by using a multichannel scaler (MCS, Laboratory Equipment, LN6500R). An Auger–photoelectron coincidence signal of about 100 counts was observed at -64 ns in an accumulation time of 120 s. The ratio of the coincidence signal to the background (S/B) was about 10:1. Since the S/B ratio of the former APECS apparatus was about 1:1 at the best [62–66], it is improved by a factor of 10 in this apparatus. Figure 16.18 shows an APECS and a singles electron spectrum of a clean Si(111) surface measured simultaneously. The APECS reflects pure Auger processes while the singles spectrum contains components of secondary electrons and photoelectrons.

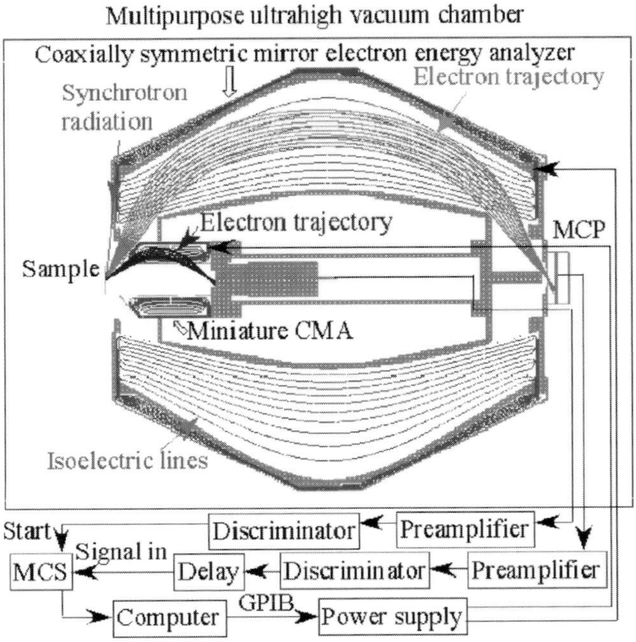

Multipurpose ultrahigh vacuum chamber

Figure 16.15: Cross section of the Auger–photoelectron coincidence analyzer [56], which consists of the miniature CMA (Figure 16.12) and a coaxially symmetric electron energy analyzer (Figure 16.2) [46]. The isoelectric lines with 10 V step, trajectories of electrons with a kinetic energy of 168 eV for polar angles of $29° - 41°$ and trajectories of electrons with a kinetic energy of 182 eV for polar angles of $51.7° - 67.5°$ are also shown based on a simulation with the SIMION 3D version 7.0. The voltages of the sample, the inner electrodes of the CMA and the coaxially symmetric mirror analyzer, and the entrance of the electron multiplier are 0 V while those of the outer electrodes of the CMA and the coaxially symmetric mirror analyzer are -100 V. A schematic diagram of the measurement system is also shown. The abbreviations used are as follows: MCP, microchannel plates; MCS, multichannel scaler.

16.5   Conclusions

We have developed a coaxially symmetric mirror electron energy analyzer, a miniature polar-angle-resolved TOF-MS, and a miniature CMA for coincidence studies of surfaces. The coaxially symmetric mirror analyzer with compensation electrodes is an ideal electron energy analyzer for coincidence studies, because it has a large transmission, a high resolution, and a high surface sensitivity. By assembling a miniature polar-angle-resolved TOF-MS in the analyzer coaxially we have developed an electron–polar-angle-resolved ion coincidence apparatus, which offers information on the yield, the mass, the desorption polar angle, and the kinetic energy of ions for the selected photoemission-final-states or the selected Auger-final-states. By assembling a miniature CMA in the coaxially symmetric mirror analyzer confocally we have developed a new apparatus for APECS which has the following features:

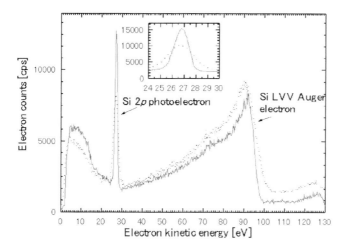

Figure 16.16: Photoelectron and Auger electron spectra of a clean Si(111) surface simultaneously measured by the coaxially symmetric electron energy analyzer (solid line) and the miniature CMA (dotted line) at $h\nu = 130$ eV [56]. The inlet shows the expanded Si 2p photoelectron
spectra.

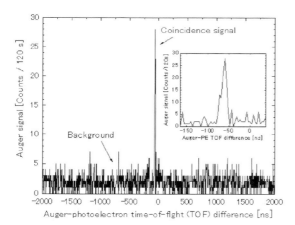

Figure 16.17: Time-of-flight (TOF) difference spectrum of Si LVV Auger started by an Si 2p
photoelectron for a Si(111) surface measured with the Auger–photoelectron coincidence apparatus at $h\nu = 130$ eV [56]. The Si 2p photoelectrons with a kinetic energy of 26.9 eV are
detected by the coaxially symmetric mirror electron energy analyzer, while Si LVV Auger electrons with a kinetic energy of 91 eV are detected by the miniature CMA. The TOF of the Si 2p
photoelectron is $65-79$ ns while that of Si LVV Auger is $7-9$ ns. The photoelectron and Auger
electron counts are 2094 ± 33 and 1677 ± 41 cps, respectively. The accumulation time is 120 s.
The Auger–photoelectron coincidence count is 103 ± 10 (0.86 ± 0.08 cps). The inlet shows the
expanded coincidence signal. PE stands for photoelectron.

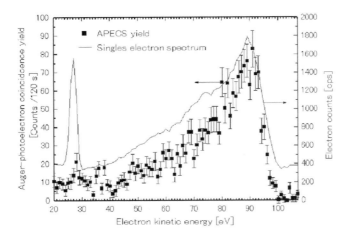

Figure 16.18: Squares denote the Auger–photoelectron coincidence spectrum of Si(111) measured by the Auger–photoelectron coincidence apparatus at $h\nu = 130$ eV [56]. The accumulation time is 120 s. The solid line refers to the single electron spectrum collected at the same time.

1) Since the solid angle of the analyzers is large, the detection efficiency of the coincidence signal is high. 2) Positioning of the two analyzers is quite easy. 3) It can be installed in a multi-purpose ultrahigh vacuum chamber. 4) The production cost is low. The most attractive point of these coincidence apparatuses is their potential. Siegbahn et al. have achieved high energy resolution of $E/\Delta E = 200-4000$ (FWHM) for solid angles of $0.94-0.19$ sr with their single-pass coaxially symmetric mirror electron energy analyzer [57]. Furthermore, the resolution can be doubled for a double-pass analyzer [57]. So, much higher resolution is expected for coincidence spectroscopy in the future. Resolution of the miniature polar-angle-resolved TOF-MS and the miniature CMA also have room to improvement. A combination of a coaxially symmetric mirror analyzer and a miniature CMA can be used for electron–energy-selected-ion coincidence spectroscopy, which will offer information on the potential energy surface responsible for ion desorption [9]. These apparatuses will bring a breakthrough in the coincidence studies of surfaces.

Acknowledgments

We express our gratitude to Mr. Y. Kobayashi, Mr. S. Terashima (KEK, Mechanical Engineering Center), Mr. T. Watanabe, Mr. M. Mori (Chiba Univ.), Mr. H. Arai (Gunma Univ.), and H. Tsutie (Hiroshima Univ.) for their valuable support in the construction and evaluation of the EICO and APECS apparatuses. We are grateful to Prof. K. Tanaka (Hiroshima Univ.), Prof. N. Ueno, Prof. K. Okudaira (Chiba Univ.), Prof. M. Okusawa (Gunma Univ.), and Prof. S. Nagaoka (Ehime Univ.) for their hearty encouragement. We also express our sincere thanks to the members of the UVSOR and PF synchrotron radiation facilities for their valuable help during the course of the experiments. This work was supported by Research for the Future Program (Photoscience; Molecular Knife – Control of Chemical Reactions by

Core Excitation, JSPS-RFTF98P01202) of Japan Society for the Promotion of Science, and by Grant-in-Aids for Scientific Research on the Priority Area "Manipulation of Atoms and Molecules by Electronic Excitation" 11222206, 12555007, and 14540314 from the Japanese Ministry of Education, Culture, Sports, Science and Technology.

References

[1] *DIET-I–V*, Springer, Berlin, 1983-1993; *DIET-VI*, Nucl. Instrum. Methods Phys. Res. B **101** (1995) 1; *DIET-VII*, Surf. Sci. **390** (1997) 1; *DIET-VIII*, Surf. Sci. **451** (2000) 1; *DIET-IV*, Surf. Sci. **528** (2003) 1.

[2] T. E. Madey, Surf. Sci. **299/300** (1994) 824.

[3] R. D. Ramsier and J. T. Yates, Jr., Surf. Sci. Rep. **12** (1991) 243.

[4] V. Rehn and R. A. Rosenberg, in *Synchrotron Radiation Research: Advances in Surface and Interface Science*, Vol. 1, ed. R. Z. Bachrach, Plenum, New York, 1992, p. 327.

[5] R. A. Rosenberg and V. Rehn, in *Synchrotron Radiation Research: Advances in Surface and Interface Science*, Vol. 2, ed. R. Z. Bachrach, Plenum Press, New York, 1992, p. 267.

[6] R. Treichler, W. Wurth, W. Riedl, P. Feulner, and D. Menzel, Chem. Phys. **153** (1991) 259.

[7] R. A. Rosenberg and S. P. Frigo, in *Chemical Applications of Synchrotron Radiation Part I: Dynamics and VUV Spectroscopy*, ed. T. -K. Sham, World Scientific, Singapore, 2002, p. 462.

[8] K. Mase, S. Tanaka, S. Nagaoka, and T. Urisu, Surf. Sci. **451** (2000) 143.

[9] K. Mase, M. Nagasono, S. Tanaka, T. Sekitani, and S. Nagaoka, Fizika Nizkikh Temperatur **29** (2003) 321 (Low Temperature Physics **29** (2003) 243).

[10] M. L. Knotek and J. W. Rabalais, in *Desorption Induced by Electronic Transitions, DIET-II*, Vol. 4 of Springer Series in Surface Sciences, eds. W. Brenig and D. Menzel, Springer-Verlag, Berlin, 1985, p. 77.

[11] K. Mase, M. Nagasono, T. Urisu, and Y. Murata, Bull. Chem. Soc. Jpn. **69** (1996) 1829.

[12] K. Mase, M. Nagasono, S. Tanaka, M. Kamada, T. Urisu, and Y. Murata, Rev. Sci. Instrum. **68** (1997) 1703.

[13] K. Mase and S. Tanaka, Jpn. J. Appl. Phys. Suppl. **38-1** (1999) 233.

[14] K. Isari, E. Kobayashi, K. Mase, and K. Tanaka, Surf. Sci. **528** (2003) 261.

[15] M. Nagasono, K. Mase, and T. Urisu, Surf. Sci. **363** (1996) 342.

[16] K. Mase, M. Nagasono, S. Tanaka, T. Urisu, E. Ikenaga, T. Sekitani, and K. Tanaka, Surf. Sci. **390** (1997) 97.

[17] M. Nagasono, K. Mase, S. Tanaka, and T. Urisu, Chem. Phys. Lett. **298** (1998) 141.

[18] K. Mase, M. Nagasono, S. Tanaka, T. Urisu, E. Ikenaga, T. Sekitani, and K. Tanaka, J. Chem. Phys. **108** (1998) 6550.

[19] K. Mase, M. Nagasono, and S. Tanaka, J. Vac. Soc. Jpn. **42** (1999) 84 (in Japanese).

[20] S. Tanaka, K. Mase, M. Nagasono, S. Nagaoka, M. Kamada, E. Ikenaga, T. Sekitani, and K. Tanaka, Jpn. J. Appl. Phys. **39** (2000) 4489.

[21] S. Tanaka and K. Mase, J. Surf. Sci. Soc. Jpn. **23** (2002) 753 (in Japanese).

[22] M. Nagasono, K. Mase, S. Tanaka, and T. Urisu, Surf. Sci. **377-379** (1997) 380.

[23] M. Nagasono, K. Mase, S. Tanaka, and T. Urisu, Surf. Sci. **390** (1997) 102.

[24] M. Nagasono, K. Mase, S. Tanaka, and T. Urisu, Jpn. J. Appl. Phys. Suppl. **38-1** (1999) 325.

[25] T. Sekitani, E. Ikenaga, K. Tanaka, K. Mase, M. Nagasono, S. Tanaka, and T. Urisu, Surf. Sci. **390** (1997) 107.

[26] T. Sekitani, E. Ikenaga, H. Matsuo, S. Tanaka, K. Mase, and K. Tanaka, J. Electron Spectrosc. Relat. Phenom. **88-91** (1998) 831.

[27] T. Sekitani, E. Ikenaga, K. Fujii, K. Mase, N. Ueno, and K. Tanaka, J. Electron Spectrosc. Relat. Phenom. **101-103** (1999) 135.

[28] K. Mase, M. Nagasono, S. Tanaka, T. Urisu, and S. Nagaoka, Surf. Sci. **377-379** (1997) 376.

[29] I. Shimoyama, T. Mochida, Y. Otsuki, H. Horiuchi, S. Saijyo, K. Nakagawa, M. Nagasono, S. Tanaka, and K. Mase, J. Electron Spectrosc. Relat. Phenom. **88-91** (1998) 793.

[30] S. Nagaoka, K. Mase, M. Nagasono, S. Tanaka, T. Urisu, and J. Ohshita, J. Chem. Phys. **107** (1997) 10751.

[31] S. Nagaoka, K. Mase, and I. Koyano, Trends Chem. Phys. **6** (1997) 1.

[32] S. Nagaoka, K. Mase, M. Nagasono, S. Tanaka, T. Urisu, J. Ohshita, and U. Nagashima, Chem. Phys. **249** (1999) 15.

[33] S. Nagaoka, S. Tanaka, and K. Mase, J. Phys. Chem. B **105** (2001) 1554.

[34] S. Nagaoka, K. Mase, A. Nakamura, M. Nagao, J. Yoshinobu, and S. Tanaka, J. Chem. Phys. **117** (2002) 3961.

[35] S. Nagaoka, J. Vac. Soc. Jpn. **46** (2003) 3 (in Japanese).

[36] E. Ikenaga, K. Isari, K. Kudara, Y. Yasui, S. A. Sardar, S. Wada, T. Sekitani, K. Tanaka, K. Mase, and S. Tanaka, J. Chem. Phys. **114** (2001) 2751.

[37] E. Ikenaga, K. Kudara, K. Kusaba, K. Isari, S. A. Sardar, S. Wada, K. Mase, T. Sekitani, and K. Tanaka, J. Electron Spectrosc. Relat. Phenom. **114-116** (2001) 585.

[38] K. Tanaka, E. O. Sako, E. Ikenaga, K. Isari, S. A. Sardar, S. Wada, T. Sekitani, K. Mase, and N. Ueno, J. Electron Spectrosc. Relat. Phenom. **119** (2001) 255.

[39] T. Sekitani, S. Wada, and K. Tanaka, J. Vac. Soc. Jpn. **46** (2003) 9 (in Japanese).

[40] E. Kobayashi, K. Isari, and K. Mase, Surf. Sci. **528** (2003) 255.

[41] S. Tanaka, K. Mase, M. Nagasono, and M. Kamada, Surf. Sci. **390** (1997) 204.

[42] S. Tanaka, K. Mase, S. Nagaoka, M. Nagasono, and M. Kamada, J. Chem. Phys. **117** (2002) 4479.

[43] S. Tanaka, K. Mase M. Nagasono, and M. Kamada, J. Electron Spectrosc. Relat. Phenom. **92** (1998) 119.

[44] S. Tanaka, K. Mase, M. Nagasono, S. Nagaoka, and M. Kamada, Surf. Sci. **451** (2000) 182.

[45] S. Tanaka, J. Vac. Soc. Jpn. **46** (2003) 20 (in Japanese).

[46] K. Isari, H. Yoshida, T. Gejo, E. Kobayashi, K. Mase, S. Nagaoka, K. Tanaka, J. Vac. Soc. Jpn. **46** (2003) 377 (in Japanese).

[47] E. Kobayashi, K. Isari, M. Mori, K. Mase, K. Tanaka, K. Okudaira, and N. Ueno, J. Vac. Soc. Jpn., **47** (2004) 14.

[48] S. M. Thurgate, J. Electron Spectrosc. Relat. Phenom. **81** (1996) 1.

[49] S. M. Thurgate, J. Electron Spectrosc. Relat. Phenom. **100** (1999) 161.

[50] M. Ohno, J. Electron Spectrosc. Relat. Phenom. **104** (1999) 109.

[51] G. Stefani, S. Iacobucci, A. Ruocco and R. Gotter, J. Electron Spectrosc. Relat. Phenom. **127** (2002) 1.

[52] M. L. Knotek and P. J. Feibelman, Phys. Rev. Lett. **40** (1978) 964.

[53] A. Kotani and Y. Toyozawa, J. Phys. Soc. Jpn. **37** (1974) 912.

[54] T. Uozumi, J. Vac. Soc. Jpn. **46** (2003) 25 (in Japanese).

[55] O. Takahashi, J. Vac. Soc. Jpn. **46** (2003) 15 (in Japanese).

[56] K. Mase, E. Kobayashi, M. Mori, Y. Kobayashi, S. Terashima, K. Okudaira, N. Ueno, J. Vac. Soc. Jpn. **47** (2004), in press.

[57] K. Siegbahn, N. Kholine, and G. Golikov, Nucl. Instrum. Methods Phys. Res. A **384** (1997) 563.

[58] W. B. Brinckerhoff, T. J. Cornish, R. W. McEntire, A. F. Cheng and R. C. Benson Acta Astronautica **52** (2003) 397.

[59] D. Coulman, A. Puschmann, U. Höfer, H. -P. Steinrück, W. Wurth, P. Feulner, and D. Menzel, J. Chem. Phys. **93** (1990) 58.

[60] K. Isari, Doctoral Dissertation, Department of Physical Science, Hiroshima University (2003).

[61] E. Kobayashi, K. Isari, M. Mori, and K. Mase, to be published.

[62] H. W. Haak, G. A. Sawatzky, and T. D. Thomas, Phys. Rev. Lett. **41** (1978) 1825.

[63] H. W. Haak, G. A. Sawatzky, L. Ungier, J. K. Gimzewski, and T. D. Thomas, Rev. Sci. Instrum. **55** (1984) 696.

[64] R. Gotter, A. Ruocco, A. Morgante, D. Cvetko, L. Floreano, F. Tommasini, and G. Stefani, Nucl. Instrum. Methods Phys. Res. A **467–468** (2001) 1468.

[65] E. Jensen, R. A. Bartynski, S. L. Hulbert, and E. D. Johnson, Rev. Sci. Instrum. **63** (1992) 3013.

[66] S. M. Thurgate, B. Todd, B. Lohmann, and A. Stelbovics, Rev. Sci. Instrum. **61** (1990) 3733.

[67] C. M. Teodorescu, D. Gravel, E. Rühl, T. J. McAvoy, Jaewu Choi, D. Pugmire, P. Pribil, J. Loos, and P. A. Dowben, Rev. Sci. Instrum. **69** (1998) 3805.

[68] K. Grzelakowski, K. L. Man, and M. S. Altman, Rev. Sci. Instrum. **72** (2001) 3362.

[69] K. Mase, E. Kobayashi, Y. Kobayashi, S. Terashima, Jpn. Pat., applied for, No. 2003-314462, Sep. 5, 2003.

Color Figures

Chapter 3

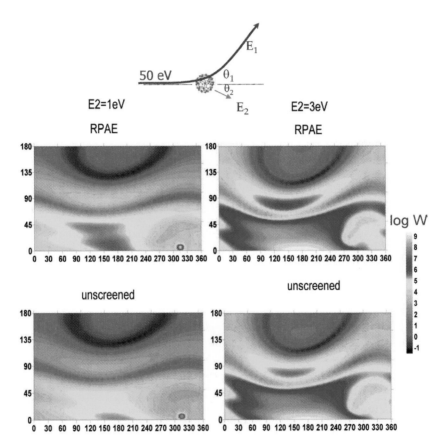

Figure 3.1: The angular dependence of the fully differential cross section for the emission of one electron from C_{60} with 1 eV (left panel) or 3 eV (right panel) following the impact of 50 eV electrons. A schematic of the scattering geometry is depicted. The upper part of the figure show the RPAE calculation while the lower part shows the calculations without treating screening effects.

Correlation Spectroscopy of Surfaces, Thin Films, and Nanostructures. Edited by Jamal Berakdar, Jürgen Kirschner
Copyright © 2004 Wiley-VCH Verlag GmbH & Co. KGaA, Weinheim
ISBN: 3-527-40477-5

Chapter 4

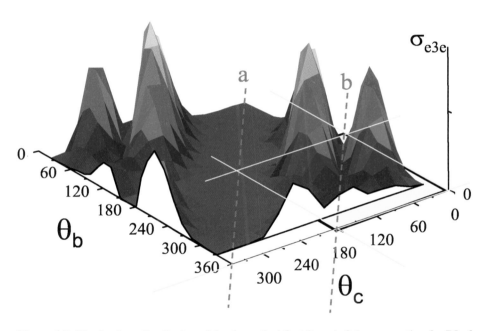

Figure 4.7: The in-plane distribution of the theoretical first Born (e,3e) cross section for DI of He, plotted *versus* both ejection angles θ_b and θ_c. Kinematical parameters as in [17]: $E_0 = 5.6$ keV, $E_b = E_c = 10$ eV, $\theta_a = 0.45°$ and $K = 0.24$ au. Unpublished results from El Mkhanter [25].

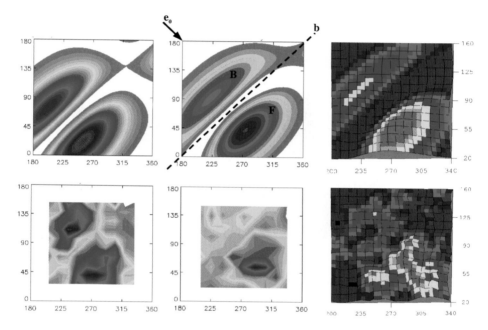

Figure 4.8: The in-plane distribution of the (e,3e) cross section for DI of He, plotted *versus* both ejection angles θ_b and θ_c. Left column: $E_0 = 5.6$ keV, $E_b = E_c = 10$ eV, $K = 0.24$ au. Middle column: $E_0 = 1.1$ keV, $E_b = E_c = 10$ eV, $K = 0.45$ au. Right column: $E_0 = 0.6$ keV, $E_b = E_c = 11$ eV, $K = 0.61$ au. Upper row: theoretical first Born CCC results, courtesy of A. Kheifets. Lower row: experiments from the Orsay group. The arrow e_0 indicates the incident electron direction, and the dashed line marked b corresponds to the back-to-back emission.

Chapter 5

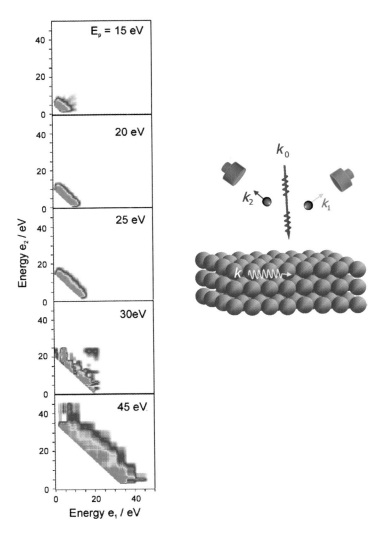

Figure 5.1: A schematic representation of the (e,2e) experimental setup in the reflection mode geometry is shown on the right hand side. The left part shows a series of energy-distribution spectra of the electron pairs at various impact primary energies shown on the figures ($E_0 = 15$ eV, 20 eV, 25 eV, 30 eV) and $E_0 = 45$ eV. The incoming electron wave vector is perpendicular the surface (a Cu(001) surface) and the two electrons are emitted symmetrically to the left and to the right of the surface normal under an angle of $40°$. The absolute values of the (e,2e) emission probability are not determined experimentally.

Chapter 10

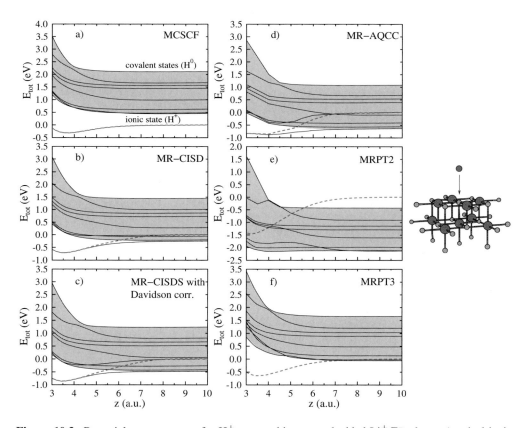

Figure 10.3: Potential energy curves for H^+ approaching an embedded $Li_{26}^+F_9^-$ cluster (vertical incidence, touch-down on Li site). Comparison of different levels of approximation: a) MCSCF, b) MR-CISD, c) MR-CISD with Davidson correction, d) MR-AQCC, e) MRPT2, f) MRPT3. The absolute energy scale is chosen such that the energy of the ionic state at large distance is 0. The dashed line indicates the *diabatic* energy curve corresponding to the ionic configuration.

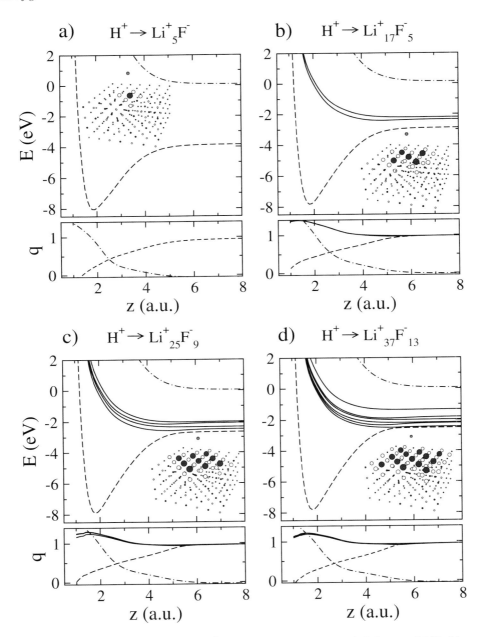

Figure 10.4: Potential energy curves for a H$^+$ ion interacting with embedded clusters of LiF of increasing size: a) Li$_5^+$F$^-$, b) Li$_{17}^+$F$_5^-$, c) Li$_{25}^+$F$_9^-$, d) Li$_{37}^+$F$_{13}^-$. Only curves of A$_1$ symmetry (within the C$_{4v}$ symmetry group) are displayed. The insets display the clusters embedded into a lattice of point charges (black: F$^-$ ions, white: Li$^+$ ions). In the lower panels below we show the electronic charge state q of the projectile. Dashed-dotted line: ionic state; dashed line: covalent state strongly interacting with ionic state; solid lines: residual states of A_1 symmetry.

Chapter 11

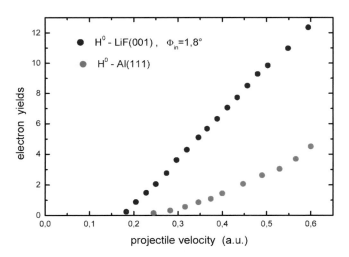

Figure 11.1: Total electron emission yields as function of projectile velocity for scattering of hydrogen atoms from LiF(001) (blue full circles) and Al(111) (red full circles) under a grazing angle of incidence $\Phi_{in} = 1.8$ deg.

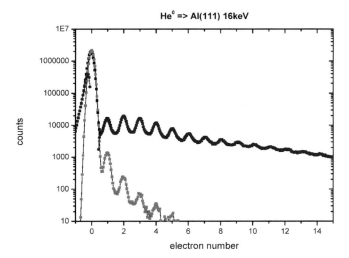

Figure 11.4: Non-coincident pulse height spectra of SBD referred to electron number for scattering of 16 keV He^o from Al(111) under $\Phi_{in} = 1.9$ deg (blue full circles). Red full squares: background signal.

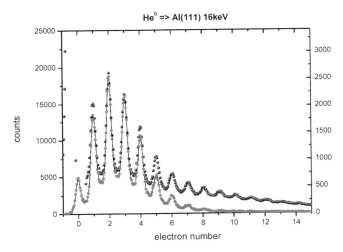

Figure 11.5: Coincident (full red circles) and non-coincident (full blue circles) electron number spectra for scattering of 16 keV He^o from Al(111) under $\Phi_{in} = 1.9$ deg.

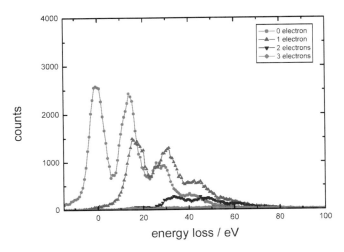

Figure 11.6: TOF spectra (converted to energy loss) coincident with number of emitted electrons for scattering of 1 keV hydrogen atoms from LiF(001) under $\Phi_{in} = 1.8$ deg. For details see text.

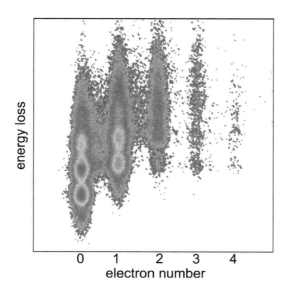

Figure 11.7: 2D plot of coincident TOF (vertical axis, "energy loss") and SBD spectra (horizontal axis, "electron number") for scattering of 1 keV hydrogen atoms from LiF(001) under $\Phi_{in} = 1.8$ deg.

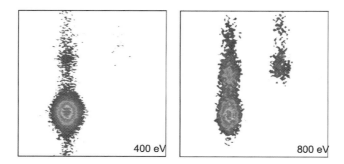

Figure 11.9: 2D plot of coincident TOF and SBD spectra for scattering of 400 eV (left) and 800 eV (right) hydrogen atoms from LiF(001) under $\Phi_{in} = 1.8$ deg.

Index

RETURN TO: CHEMISTRY LIBRARY
100 Hildebrand Hall • 510-642-3753

LOAN PERIOD 1	2 1 Month	3
4	5	6

ALL BOOKS MAY BE RECALLED AFTER 7 DAYS.
Renewals may be requested by phone or, using GLADIS,
type **inv** followed by your patron ID number.

DUE AS STAMPED BELOW.

MAY 2 0 2005		
DEC 1 6		
Feb 25th 2009		
OCT 0 4 2009		

FORM NO. DD 10
3M 5-04

UNIVERSITY OF CALIFORNIA, BERKELEY
Berkeley, California 94720–6000